*The Political Economy of Global Energy*

Rajendra K. Pachauri, Director of the Tata Energy Research Institute at New Delhi, is an energy economist and author of five previous books on the subject. He wrote *The Political Economy of Global Energy* when he was professor of resource economics at West Virginia University.

# THE POLITICAL ECONOMY OF GLOBAL ENERGY

R. K. PACHAURI

The Johns Hopkins University Press
BALTIMORE AND LONDON

© 1985 by The Johns Hopkins University Press
All rights reserved
Printed in the United States of America

The Johns Hopkins University Press, 701 West 40th Street, Baltimore,
Maryland 21211   The Johns Hopkins Press Ltd, London

**Library of Congress Cataloging in Publication Data**

Pachauri, R. K.
The political economy of global energy.

Bibliography: p.
Includes index.
1. Petroleum industry and trade.  2. Energy industries.
3. Energy consumption.  4. International economic relations.  I. Title.
HD9560.5.P23   1985        333.79        84-21825
ISBN 0-8018-2469-9
ISBN 0-8018-2501-6 (pbk.)

Typeset by Taj Services Ltd, India

# CONTENTS

List of Figures vii
List of Tables ix
Acknowledgments xiii
Introduction 1

1. Global Energy, OPEC, and International Economic Relations 5
2. Global Energy Transition 12
3. Future Energy Prospects 30
4. The History and Evolution of OPEC 53
5. Oil Revenues: Determinants and Impact 76
6. OPEC and the Global Economy 94
7. The International Economic Order 115
8. Agenda for Change 129
9. Future Scenarios and Perspectives 149
10. Conclusions 169

Bibliography 179
Index 189

# FIGURES

1.1  Global Energy Economy Interactions   *9*
2.1  Estimated International Recoverable Reserves of Coal, 1978   *20*
2.2  International Coal Production, 1979   *22*
2.3  International Hydroelectric Power Production   *23*
2.4  Rate of Discovery of World Oil Reserves   *24*
2.5  International Petroleum Supply and Disposition, 1978   *27*
3.1  Free World Energy Consumption   *41*
3.2  Free World Oil Consumption   *41*
4.1  Market Forces Acting on a Cartel   *62*
4.2  Sensitivity of OPEC to Market Changes   *66*
4.3  OPEC Stability Index, 1973–82   *68*
6.1  World-Market Commodity Prices; Oil and Nonoil   *97*
6.2  Oil-Importing Developing Countries, Ratio of Debt to Exports of Goods and Services, 1975–81   *108*
6.3  Oil-Importing Developing Countries, External Debt Service Payments, 1973–81   *109*

# TABLES

2.1 Energy Intensity Indexes, by Country Group, Selected Years, 1960–80 *13*

2.2 Index of Energy Prices to Final Users, by Country Group, 1975, 1978, 1980 *15*

2.3 Energy Consumption in Industrial Countries, Selected Years, 1960–80 *16*

2.4 Energy Consumption in Oil-Exporting Developing Countries, Selected Years, 1960–80 *17*

2.5 Energy Consumption in Oil-Importing Developing Countries, Selected Years, 1960–80 *18*

2.6 Petroleum Reserves and Production, by Country Group, 1960, 1970, 1980 *25*

2.7 World Drilling Effort, by Country Group, Selected Years, 1965–78 *25*

2.8 Commercial Primary Energy Production and Consumption, by Country Group, 1970, 1978, 1980 *28*

3.1 Average Annual Growth of Gross National Product, by Country Group, 1960–90 *37*

3.2 Commercial Energy Consumption, by Country Group, 1980 and 1990 *38*

3.3 Growth in Energy Consumption and Gross National Product, by Country Group, 1960–2000 *40*

3.4 Coal Production, by Country Group, 1970, 1980, 1990 *43*

3.5 Production of Primary Electricity, by Country Group, 1970, 1980, 1990 *45*

3.6 Known Recoverable Oil Reserves, by Country Group, End of 1981 *48*

3.7 Blinkers-on Scenario, Oil Production, Selected Years, 1980–2000 *50*

3.8 Natural Gas Reserves by Country Group, 1970 and 1980 *50*

3.9 Projections of Primary Energy Production, by Country Group, Selected Years, 1980–2000 *51*

ix

x / Tables

4.1  Changes in Annual Oil Output Since 1973, OPEC Countries, 1975–81   *64*
5.1  OPEC Oil Revenues   *77*
5.2  Exports of Goods and Nonfactor Services, OPEC Countries   *78*
5.3  Econometric Estimates of the Composition of Gross Domestic Product, Selected OPEC Countries   *91*
5.4  Projected Rates of Annual Growth of Nonoil Gross Domestic Product, Selected OPEC Countries, 1982–2000   *93*
6.1  Changes in Consumer Prices, Industrial Countries, 1973–81   *95*
6.2  Changes in Components of Demand, Industrial Countries, 1974   *97*
6.3  Recycling of International Deficits and Surpluses, Country Groups, Selected Periods, 1950–80   *99*
6.4  Deployment of OPEC Investible Surplus, 1974–79   *100*
6.5  Return on OPEC Commercial Investments, by Country, 1976–78   *102*
6.6  Balance on Current Account, Developing Countries, Selected Years, 1973–81   *106*
6.7  Growth Rates of OECD and Developing Countries, 1973–82   *107*
6.8  Bank Debts of Developing Countries, June 1982   *107*
6.9  Flows between Bank of International Settlement Reporting Banks and Country Groups outside Reporting Area, 1974–80   *110*
7.1  Effect of Changes in Terms of Trade and Export Volume Oil-Importing Developing Countries, 1974–78   *116*
7.2  Merchandise Trade, Oil-Importing Countries, 1970 and 1980   *122*
7.3  Composition and Growth of World Merchandise Trade, 1970–80   *122*
7.4  Share of Merchandise Exports, by Country Group, 1960 and 1980   *123*
7.5  Composition of Trade in Manufactures, Developing Countries, 1979   *125*
8.1  Current-Account Deficit and Finance Sources, Oil-Importing Developing Countries, Selected Years, 1970–80   *134*
8.2  Borrowing in the Medium-Term Eurocurrency Credit Market, 1973–81   *135*
8.3  Terms for Syndicated Eurocredits, 1975–1981   *136*
8.4  Classification of Developing Countries, by Oil Imports and Fuelwood Scarcity   *139*
8.5  Official Assistance to Developing Countries, OPEC, OAPEC, and OECD Countries, 1969–82   *143*
8.6  Institutions Distributing OPEC Aid   *144*
8.7  Remittances of Migrant Workers in Middle East Countries to Their Home Country, as Ratio of Home Country Merchandise Exports   *146*

*Tables* / xi

9.1 Oil Price Assumptions Underlying Scenarios for the Future, Middle East Capital Surplus OPEC Countries  *150*
9.2 Projected Oil Production, Consumption, and Exports for Capital Surplus Middle East OPEC Countries  *150*
9.3 Nonoil Gross Domestic Product, Middle East Capital Surplus OPEC Countries, 1974–2000  *151*
9.4 Blinkers-on Scenario, Nonoil Gross Domestic Product, Middle East Capital-Surplus OPEC Countries, 1985–2000  *152*
9.5 Iraq: Blinkers-on Scenario, Price Assumption $y$, 1981–2000  *153*
9.6 Kuwait: Blinkers-on Scenario, Price Assumption $y$, 1981–2000  *154*
9.7 Libya: Blinkers-on Scenario, Price Assumption $y$, 1981–2000  *155*
9.8 Qatar: Blinkers-on Scenario, 1981–2000  *156*
9.9 Saudi Arabia: Blinkers-on Scenario, Price Assumption $y$, 1981–2000  *157*
9.10 United Arab Emirates: Blinkers-on Scenario, Price Assumption $y$, 1981–2000  *158*
9.11 Blinkers-on Scenario, Middle East Capital Surplus OPEC Countries, Selected Years, 1981–2000  *158*
9.12 Assumptions Underlying Scarcity Scenario, Middle East Capital Surplus OPEC Countries, Selected Years, 1980–2000  *159*
9.13 Iraq: Scarcity Scenario, 1981–2000  *160*
9.14 Kuwait: Scarcity Scenario, 1981–2000  *161*
9.15 Libya: Scarcity Scenario, 1981–2000  *162*
9.16 Qatar: Scarcity Scenario, 1981–2000  *163*
9.17 Saudi Arabia: Scarcity Scenario, 1981–2000  *164*
9.18 United Arab Emirates: Scarcity Scenario, 1981–2000  *165*
9.19 Scarcity Scenario, Price Assumption $x$, Middle East Capital Surplus OPEC Countries, 1981–2000  *166*
9.20 Scarcity Scenario, Price Assumption $y$, Middle East Capital Surplus OPEC Countries, 1981–2000  *167*

# ACKNOWLEDGMENTS

The author would like to express his deep gratitude to all those who provided help and encouragement in various ways in the completion of this book. Valuable comments and suggestions on various chapters of the book were provided by Professor William Miernyk, Dr. Fereidun Fesharaki, and Dr. James Paddock. Considerable assistance, encouragement, and support were given by Professor Adam Rose, Dr. Harrison Brown, and Dr. Kirk Smith.

It would not have been possible for the author to complete this work without the valuable assistance provided by Rashmi Pachauri and Leena Srivastava, particularly in the computational and quantitative work presented in this book. The author is deeply grateful to them for this and for their all-round assistance.

My deep thanks are also due to Alice Kerns for typing the first draft of the manuscript in record time, to Anupam Chopra for typing a subsequent version, and to B. Sundararaman for typing the final version.

Last but not least, my deep appreciation and thanks are due to my wife, Saroj, to my daughters, Shona and Moneesha, to my brother, Dr. S. K. Pachauri, and to my parents, for encouragement, inspiration, and understanding.

*The Political Economy of Global Energy*

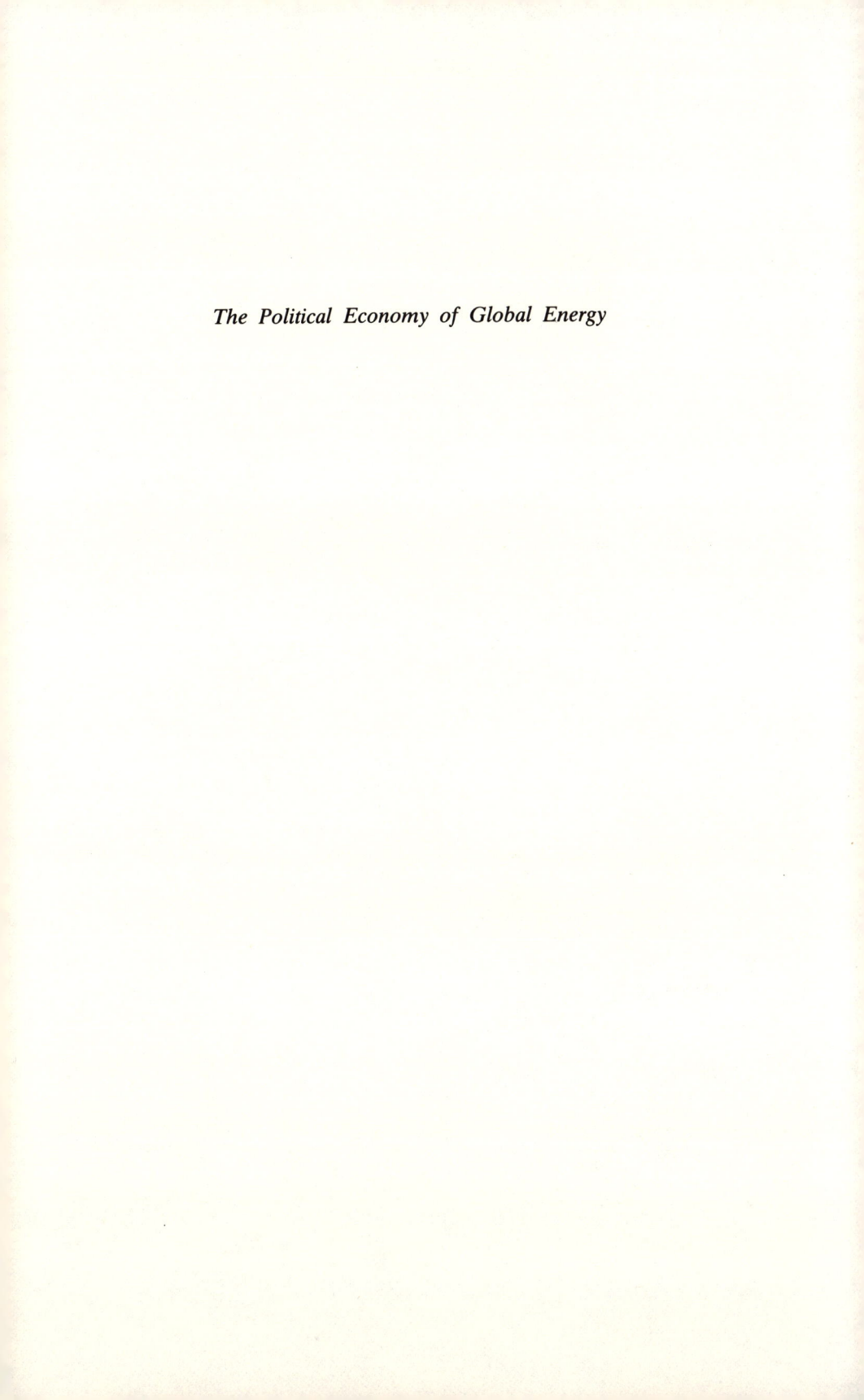

# INTRODUCTION

Major changes have taken place in the world energy economy during the decade between 1971 and 1981. Two reports published in early 1983 indicate that the total demand for energy in the countries of the Organisation for Economic Cooperation and Development (OECD) grew only 15.7 percent during the period 1971–81 as against approximately 90 percent in the previous decade.[1] Also, this lower rate of growth was not uniformly distributed over the various sources of energy. For instance, oil demand grew only 3.6 percent during the period, coal 22.4 percent, and natural gas 12.9 percent. Hydropower increased by 26.9 percent, and nuclear power, starting from a low base, grew by almost 600 percent, reaching a share of 5 percent in total energy consumed. Oil consumption was cut back significantly in the industrial and residential sectors; and even though demand in the transport sector did go up in this period, it was at a slower rate than during the previous decade.

While these interfuel and intersectoral changes in consumption were taking place, major changes occurred in the structure of the energy supply industry as well. Supply-side changes, in fact, were largely responsible for bringing about changes in the consumption of energy. In particular, the emergence of the Organisation of Petroleum Exporting Countries (OPEC) as a powerful body, influencing the quantities and prices of oil in the world market, has accounted for unprecedented developments in the global economy. Yet OPEC's fortunes have fluctuated substantially. The glut in the oil market since 1979–80, followed by the initial softening and subsequent decline in world oil prices, seems a far cry from the shortages of gasoline in the fall and winter of 1973.

Questions are inevitably being raised about the stability of prices in the future, the continuance of OPEC as an effective market force, and

---

1. International Energy Agency, *Energy Statistics of OECD Countries*, (Paris: OECD); and International Energy Agency, *Energy Balances of OECD Countries* (Paris: OECD).

the future demand for oil by the noncommunist nations, given the recession of the early 1980s and the results of worldwide conservation efforts. Economists, futurologists, and crystal-ball gazers have produced a variety of scenarios related to these issues. In mid-1983, members of the International Association of Energy Economists (IAEE) at their annual North American conference pondered these questions and focussed on a number of recent observations ("Energy Economists Unite" 1983). The year 1983 was seen as a period of no growth in the demand for oil in the noncommunist nations and a possible 1 percent growth in total demand for energy. Oil was, however, predicted to bounce back as part of the expected 4 percent growth in energy demand during 1984. Consequently, it was projected that OPEC would raise its production quotas during late 1983. The usable commercial inventory of petroleum, which had risen to twenty-seven days of forward demand by the second quarter of 1981, was seen to be around ten days in mid-1983. (This figure is not far from the all-time low of six days on April 1, 1979.) Like the deliberations of the IAEE, most recent energy literature published in journals and trade magazines is concentrating on short-run changes in the world oil market. Undoubtedly these issues are important, particularly if one considers the possibility of violent market fluctuations, which could break OPEC's cohesiveness (much as oil prices broke expected ceilings in the eventful periods of 1973–74 and 1979–80).

There are, today, very tentative and guarded forecasts on the long-term prospects of the world's oil market conditions. Most of them are optimistic and at least dampen earlier forecasts of steep increases in the real prices of energy through the rest of the century. The World Bank, for instance, which had earlier forecast a real-price increase of between 2 and 3 percent a year, has now projected that in the mid-1990s the real price of oil will be around 20 percent above its 1981 peak. The central scenario developed by the Bank assumes that the oil price increase in real terms will average about 1.6 percent a year between 1982 and 1995. But a more important aspect of oil price increases is the implication they would have for the structure of the oil market. The critical question is whether OPEC will remain an effective force in the future. Bijan Mossavar-Rahmani and Fereidun Fesharaki predict a rebound of the exporters.[2] Their thesis is based on what they call the OPEC multiplier, which they define as the mechanism by which a swing in energy demand is magnified in the demand for OPEC oil, which is the world's residual energy source. The authors feel that the same forces

2. "OPEC and the World Oil Outlook: Rebound of the Exporters?" Special Report 140 (London: Economist Intelligence Unit, 1983).

that dramatically drove down demand for OPEC oil in the early 1980s will probably drive it up again over the next few years, thereby driving up oil prices sharply. Another aspect of their thesis is their disagreement with what is now identified as "irreversible conservation." Mossavar-Rahmani and Fesharaki contend that a number of energy developments that were anticipated immediately after the last oil shock have either been reduced, postponed, or cancelled completely. This fact combined with the recent decline in real prices for oil will result in increased demand after 1983 and a surprising rebound in the demand for OPEC oil.

Much of the dispute on this subject arises out of the differences in the perceived strength of OPEC as an effective decision-making force. The impending collapse that many observers saw in OPEC's strength and ability to hold together in early 1983 has been belied by subsequent events. Despite a reduction of 15 percent in the real price of oil and the imposition of rigid production quotas for the first time in OPEC's history, it has managed to stick together reasonably well. How changes in the future will affect its cohesiveness is a subject for much deeper analysis than is usually practicable, although some part of it has been attempted in this book. In general, the answer hinges on the ability of the member nations of OPEC to adjust successfully to significant changes in oil market conditions.

Assessments of the ability of individual OPEC nations to absorb large inflows of oil revenues vary considerably, particularly for the nations of the Middle East. These assessments have also undergone substantial change over the past ten years, as a result of changes in both the surpluses generated by the OPEC nations (which have had a profound impact on international banking activities) and the realization that has grown within these nations of the dubious benefits of accelerated investments in the nonoil sectors of their economies. A recent publication reviewing the post-1973 developments and their effects on future development strategies in the oil-exporting nations seems to suggest some major changes in the economic management of these nations.[3] The challenge ahead is particularly arduous for these nations on account of growing external debts in some of them, heavy internal development obligations in others, and growing social problems attendant on rapid economic growth in still others. These nations are also influenced by exchange rates and trade policies in the international market. Oil production and pricing efforts are, therefore, likely to be influenced heavily by both international developments and internal

3. Jahangir Amuzegar, *Oil Exporters' Economic Development in an Interdependent World* (Washington, D.C.: International Monetary Fund, 1983).

forces that favor a slowing down of developmental efforts in the coming decades.

A component of the global energy picture which has thus far received inadequate attention is the situation obtaining in the oil-importing developing countries. These nations and their actions in the future are critical to any assessment of likely global energy developments. Of particular importance is the relatively high rate of growth in energy demand in a number of Third World nations, even though the demand in the industrial nations has slowed down considerably or actually been reduced. At the same time, there are two constraints on Third World energy consumption and economic development. First, many of these countries lack the financial and other resources for indigenous energy development. Second, their adverse situation in the international economic order often leaves them unable to afford to import the quantity of energy essential to maintain standards of living and of economic activity. The clamor for changes in the international economic order is, therefore, most notably from some of these nations. But this demand is also voiced, often vociferously, by OPEC nations, since they too have a vital self-interest in stable exchange rates and concessions to counter the effects of inflation in countries that supply them with capital and consumer goods. Besides, OPEC members feel that their political moorings in the international arena are in the developing countries, so the demand for a new international economic order is as much an exercise in global geopolitics as an expression of economic self-interest.

In the following pages we explore some of these issues in the context of very recent developments. The empirical support buttressing various arguments is generally limited to the 1973–80 period, in which the effects of the first oil price shock can be observed. The effects of the second oil price shock in 1979–80 are still not clear, and therefore no long-term conclusions can be drawn. One effect of the two severe oil shocks and the intervening period (when the oil exporters held dominant sway) is what some refer to as a counterrevolution in the global oil market, with the oil-consuming nations calling the tune. It is too early to comment on the effects of the period 1980–83, but the present stability of the world oil market does provide evidence that, in the long run, the oil market will continue on the path established over the past ten years or so. There were several periods of tightening in the oil market; the period 1980–83 was a period of slackening. Analysts and observers must, therefore, look beyond the present temporary situation to see the long-term possibilities.

CHAPTER 1

# GLOBAL ENERGY, OPEC, AND INTERNATIONAL ECONOMIC RELATIONS

The economy of the world is in a period of transition that is likely to continue for many decades. At the time of writing this book, we are more than ten years into the period ushered in when a group of leaders from the Arab nations imposed a boycott on oil supplies for some nations of the West in October 1973. In the immediate aftermath of this event, panic conditions developed in the world oil market. Oil prices spiralled upward and continued to climb except for brief and deceptive periods of stability, such as those of 1977–78 and 1981–82 and the decline in nominal prices in 1983. The shift in oil market power since 1973, the outlook for higher prices in the future, and the vulnerability of oil importers to sudden supply disruptions are all responsible for the efforts to restructure economic systems over these years. Of all events of importance to the global economy in this century, the oil price revolution of the 1970s has had the most far-reaching effects.

The historic significance of this event goes beyond changes in the world's energy picture and questions about the stability of the Organisation of Petroleum Exporting Countries (OPEC). The post-1973 changes in the world oil market have in fact brought about major changes in the international banking system, triggering large-scale alterations in world capital flows, with actual transfers of major resources between countries and regions. All this has set in motion a major restructuring of economic systems, with accelerated replacements of capital stock and technological changes in consonance with higher energy prices. Most studies on this subject in recent years have, understandably, focussed mainly on bilateral issues between OPEC and the industrial nations and have generally ignored the third leg of the global triangle, namely, the oil-importing developing nations. Oil market developments have had a major impact on these countries of the Third World; and even though OPEC pricing decisions have been consistently supported by these nations as part of an overall move for higher commodity prices, they have chosen to suffer in silence the burden of higher import bills and economic hardship to serve what they consider their long-term interests.

Any prognostication of energy or global economic activity is hazardous. The oil price increase of 1979–80 induced by political changes in Iran, the subsequent recession in the western world, and the oil market glut of 1981–83 all appear to have taken the world community by surprise, despite the experiences of the mid-seventies. A measure of the uncertainty in forecasting energy and global economic developments is the rapidity with which projections have been revised (mainly downward) in recent years. The future is so clouded by uncertainties of economic recovery in the North and by short-run accelerated growth in the South that long-range projections are tentative and hazy.

At the core of these uncertainties are the questions: How rapidly will demand for energy grow in the future? How much of the energy demanded will be in the form of oil? What will be the prices of oil and of alternate sources of energy? How much oil will OPEC supply and at what price? How will capacity and production grow among non-OPEC oil producers? What levels of finance will be available to the Third World countries for development in general and production of energy in particular? How will the international economic order evolve to enhance or inhibit these developments? Answers to each of these questions are complex by themselves, but they are complicated further by the linkages that tie them together. Any forecast of the future, therefore, must be based on assessments of the nature and magnitude of the cause and effect relationships among global energy, OPEC's actions, and the world economy. Recent behavior of this entire system provides us with useful information for developing a vision and plausible scenarios of the future.

No specific forecasting models have been developed for this book; nor has a unique time path of future world energy demand been defined. I have preferred to use forecasts from other studies as a basis for discussing the issues mentioned above and for putting forward scenarios of my own. It is, of course, difficult in a study of such broad sweep to isolate developments on the demand side from those on the supply side. Increased demand brings about an increase in price, and this in turn influences the production (and pricing) decisions of major oil producers. Similarly, output restrictions and higher prices bring about reductions in quantity demanded over a period of time. Hence the scenarios of demand presented in this book take into account the scenarios of anticipated supply.

Discussions on the supply of energy emphasize the dominance of oil through at least the rest of this century and are confined to the world oil market. This is as much due to limitations of space as to the fact that oil will be the major determinant of energy developments for the next quarter century. Discussions on the oil industry explain the importance of OPEC as the critical source of supply in the foreseeable future and challenge the myth of OPEC's decline. For adopting this approach I may, perhaps with some justification, be accused of a bias in favor of the oil-exporting nations, who are often viewed as an avaricious and hostile entity exerting monopolistic power to extract huge rents and political concessions from the beleaguered consuming nations. But reality has to be faced squarely, and the case must be made that OPEC is in a strong position for reasons of history and economics and the growing scarcity of oil—predictions of redressal of the grievances of oil importers is simple self-deception. It is also insinuated that OPEC decision makers are a clumsy and unenlightened lot, who are not aware of the effects of their actions. Undoubtedly these individuals are not infallible and are even given to serious error, as the price increases of 1979–80 indicate in retrospect; but it would be naive to believe that OPEC decisions are ill informed or made on poor advice.

An appraisal of OPEC's role in the world's energy future has also to be seen in the context of the whole question of natural resources and their depletion. Before the pioneering work of Hotelling (1931) over a half century ago, economists generally ignored natural resources in their analyses and discussions of economic activity. Staunch believers in the nonclassical school continue to believe that the oil crisis of the seventies and eighties is a temporary aberration, which will be corrected with new discoveries of oil and other substitutes, and that utopia will be with us again in the shape of single-passenger gas-guzzling automobiles. Such views weaken the public's belief in the market as an efficient allocator of resources. These views ignore the growing scarcity of resources and the time lag until substitute resources and substitute lifestyles and social habits evolve.

The pain and social cost of transition cannot be wished away, for it encompasses far more than a 20 percent unemployment rate in Detroit. The transition triggered by higher oil prices has to do with the dependence of the human race on what Georgescu-Roegan (1980) calls "exosmatic," or detachable organs in the form of automobiles or planes or electric can openers (Pachauri 1980*b*). According to the bioeconomic view of human activity, the sudden limitations on the use of these exosmatic organs can be very costly in terms of human welfare. In essence, a reduced or modified use of these organs induces evolutionary change, the speed of which is determined by a variety of factors, including the rate of replacement of these devices, modification of these devices, substitute means for running these devices, the availability of substitute devices, and society's psychological attitude to change and evolution. Adherents to the philosophy of growth often fail to see that a restructuring of the economic system cannot be without cost and must deviate from the decades-old path of exponential growth of consumption. It must be remembered that this growth was fuelled by erroneous signals of plenty from a market controlled by decisions not reflective of future scarcity.

Concern with oil and its worldwide effects is related to the global nature of the transition, with its inequitable dimensions in relation to North and South. The institutional strengths of organizations and mechanisms in countries of the North facilitated considerably the first round of adjustments in the wake of the 1973-74 oil price rise. It is not clear, on the basis of existing studies, whether the South had quite the same success, particularly considering the prospects for economic development in the immediate aftermath. There is a view widely held that the second round of effects on the Third World will be very severe because of their high levels of debt in the international market, postponement of important projects, higher import bills, scarcity of external finance, and high rates of population increase. Developments in the oil market touch on all these Third World problems and are at the heart of any analysis of a stable and more equitable international economic order. Since the world today is a much smaller place than it ever was, concern for human welfare in a global perspective goes beyond altruistic or noble motives and is intimately connected with the stability of the system. A transition from one state to another can be either stable or chaotic, and the latter will result in many more conflicts like that over the Falkland Islands.

To clarify the interdependence between energy and international issues, the book focusses on linkages among global energy developments, national economic trends, and the monetary flows among the industrial nations, OPEC, and the oil-importing developing countries of the world. One reason some researchers and policymakers view OPEC decisions within a neoclassical economic framework is that OPEC's links with a variety of global developments have come into being only recently. The 1973-74 oil price increase brought into existence a totally new balance of economic and political power, which instituted international economic relationships that are still evolving. In simple form, these relationships are as shown in figure 1.1.

At the core of these new economic relationships is OPEC's decision-making structure, which is deeply influenced not only by exogenous developments, such as the effects of recession and energy

Global Energy, OPEC, International Economic Relations / 9

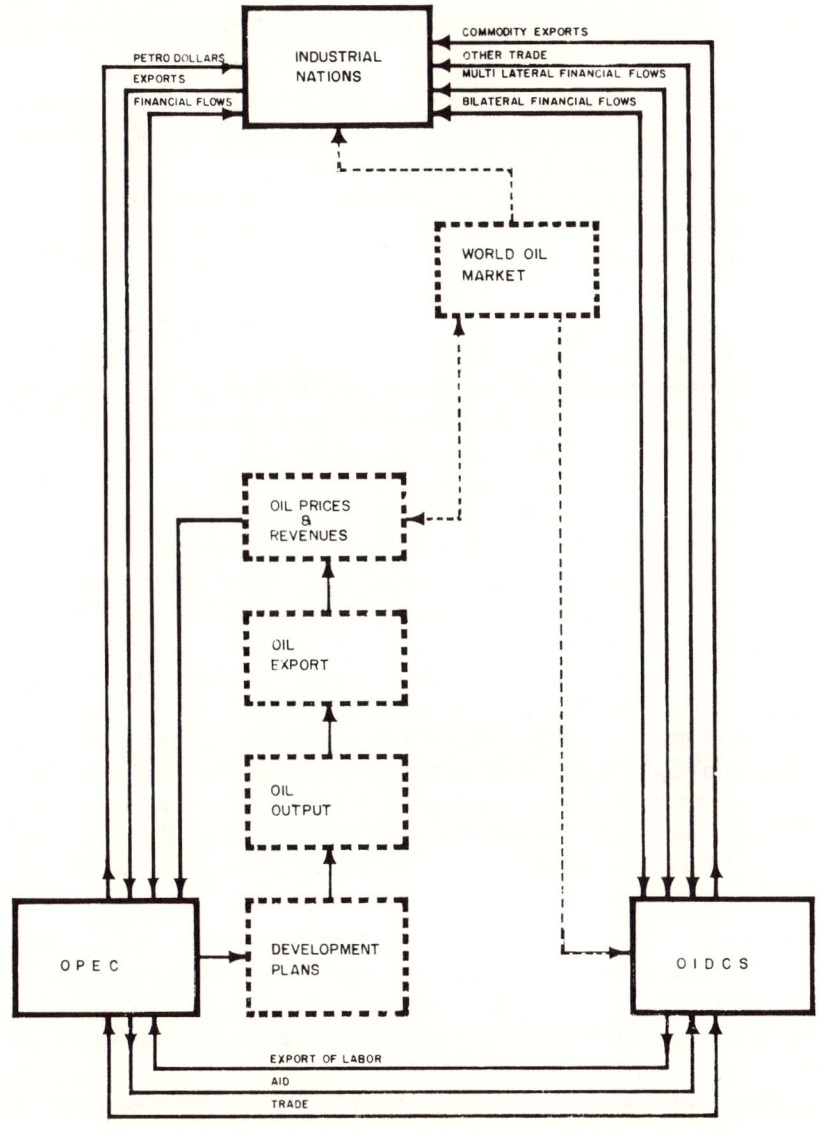

Figure 1.1 Global Energy Economy Interactions

conservation in the industrial nations, but also by the social and economic effects of oil revenues invested in the oil-exporting nations themselves. These factors are explored in detail in subsequent chapters, but the following brief summary can be made in the context of figure 1.1.

1. OPEC's output and pricing decisions, although partly a response to developments in the world oil market, are also significantly based on internal developments and plans.
2. Oil prices and quantities exported are strongly linked to the effects on society of rapid investments in the economies of oil-exporting nations, since these investments are based on factors such as the growth of infrastructure and absorptive capacity in each country and the equity effects of oil revenue earnings and investments.
3. Global monetary flows and international trade are important determinants of OPEC's oil exports, since the members of OPEC with capital surplus generally invest their surpluses in the industrial nations of the West. They are, therefore, tied into the private banking system of the world and have a direct interest in capital flows and returns from overseas lending and investments. Surpluses are also spent on goods and services, largely from the West. Thus the rate of inflation and domestic demand in the industrial nations affect trade between them and OPEC.
4. Nonoil developing countries form a part of this complex set of interrelationships by (1) supplying labor to the oil-exporting nations of the middle east; (2) borrowing capital from the private banking system, which in turn depends on deposits from those members of OPEC with capital surplus; (3) importing goods and services from the industrial nations; (4) politically supporting OPEC; and (5) pushing for the establishment of a new international economic order in concert with OPEC.

Another important aspect of this triangular relationship is the likelihood of developing countries demanding a larger share of oil from the world market in the coming decades. This increase is likely to exert pressure on the global oil market and may be a major factor determining future oil price increases.

These and related issues have been explored in a recent publication by Hoffman and Johnson (1981), which focusses on the dangers of the "triangle of suspicion" that seems to characterize current transactions among OPEC, the developed countries, and the Third World. The authors rightly feel that the energy crisis has been almost equally corrosive of trust in both the industrial North and the oil-poor South.

There is at the heart of this triangle a complex set of forces, which have caused past negotiations and debates among the three groups to lead to frustration on all sides, often ending in serious animosities. Yet on some issues there has been some identity of views between two members of these three groupings. For instance, OPEC and the developing countries have had a generally identical position vis-à-vis the industrialized North in the negotiations for the new international economic order. But in other respects, the developing countries are suspicious of OPEC, fearing the possibility of a separate deal between oil exporters and the countries of the North; they are also resentful of the crippling effects of the sudden oil price increases of 1973–74 and 1979–80 on their fragile economic systems. Also, exporters of other commodities, who once saw in the success of OPEC an inspiring example of what could be achieved through Third World unity, have now come to realize that nonoil commodities do not possess even a fraction of the leverage of oil.

The central thesis that emerges from these observations is that energy developments are interwoven with the functioning of the international economic order. Consequently, changes in the energy market, with their large-scale effects on petrodollar investments and international flows, would be expected to give rise to demands for changes in the international economic order, just as changes in this order would have an impact on the economic and political forces that determine oil market conditions. An analysis of one segment of this system must, therefore, be complemented by a study of other parts of the whole. Hence, the new set of linkages which have come into existence in the past ten years require scrutiny to determine how developments in one part of the system are transmitted to other parts of the global triangle.

This book views the global energy problem in the aggregate, analyzes the evolution of OPEC, its structure and stability, and develops scenarios for the future for world energy as well as for the international economic order. Following chapters are arranged in this general sequence and the concluding chapter describes those policies that will ensure a stable order in the next two decades. Needless to say, a futuristic study is limited by institutions and structures as they exist today, and sudden and unexpected changes in these can alter scenarios for tomorrow quite drastically. A Saudi Arabian uprising similar to the Iranian revolution could spell disaster for the world economy; the collapse of OPEC as an effective organization could send severe tremors through the entire global system. Such happenings could make nonsense of any futuristic study in the field, and this book would not be an exception.

# CHAPTER 2
# GLOBAL ENERGY TRANSITION

Global energy developments since World War II have been marked by a growing dependence on oil caused both by rates of economic growth much higher than achieved earlier and by a shift to oil from coal and other sources, largely on account of lower relative prices for oil right up to 1973. Historically, the advent of oil as a major source of energy is associated with the entry of the automobile as a form of transportation, but a shift to oil had started even earlier in other applications. As a matter of fact, in the early part of this century, fuel oil, which was a new energy source, survived competition from and then actually replaced coal, the existing fuel of the day.

The U.S. record of energy transition is very significant, since at similar stages of development various nations subsequently followed the same path. They were aided in this move by the effectiveness of the oil companies in marketing oil products at constantly declining prices all over the world. Consequently, even those nations that did not possess the advantages of indigenous oil production were overwhelmed by the sheer economic benefits of using oil and did so, even in some cases to the disadvantage of indigenous coal production. The displacement of coal, particularly in the United States, was very rapid, considering the vast amount of coal-using capital stock built up during the second half of the nineteenth century. In 1850, coal supplied only 9.3 percent of total U.S. energy (fuelwood accounting for the balance), but by 1900 this figure had increased to 71.4 percent (Schurr and Netschart 1960, 36, 145). By the turn of the century, therefore, coal was firmly established as the main source of energy in the United States, and the transport system had been built around moving coal from the Appalachian region to the major consuming centers in the Northeast and the Midwest. The rapid expansion of the U.S. economy between 1900 and 1920 resulted in a 100 percent increase in energy consumption, and the discovery of large oil fields in Texas and Oklahoma brought forth a gush of oil that captured the new markets opened up in this period. The close proximity to easy oil was an important factor in the development of California and the U.S. Southwest. There was little changeover of coal-using equipment to oil use, but oil captured a larger share merely through the design of new industries for fuel oil use. Much the same developments took place in Europe and Japan during their rapid industrialization after World War II. Subsequently, the developing countries of the Third World have followed a similar path, and their choice of oil was influenced by (1) the availability of oil-using technologies and industrial plants and the eagerness of western equipment suppliers to design for oil use; and (2) the steadily declining prices of oil during the fifties and sixties.

In any assessment of the future it is important to remember that the developing countries are at the beginning of a period of intensive industrialization, modernization of transport systems, and a shift from traditional fuels like firewood and animal and vegetable waste to commercial sources of energy. Whereas the developed countries have achieved significant reductions in energy use per unit of output, particularly since 1973, the developing countries as a group are likely to continue with higher intensities of energy use for many years to come. For instance, in the 1960s in the United States about eight and a half barrels of oil equivalent were consumed per $1,000 of gross national product in 1975 (Exxon 1979, 8). By 1978, 6 percent less energy was used per unit of output. This was equivalent to a reduction in yearly U.S. demand of around two to two and a half millions of barrels a day (mbd). In the other developed countries, the overall pre-1970 ratio was five and a quarter barrels of oil equivalent for $1,000 of gross national product, but by 1978 it had been reduced by 3 percent, resulting in an implied reduction in demand of around one to one and a half mbd of oil equivalent. These energy intensities are shown in table 2.1. While there

TABLE 2.1

Energy Intensity Indexes, by Country Group, Selected years, 1960–80
(1973 = 100)

| Country Group | 1960 | 1975 | 1978 | 1980[a] |
|---|---|---|---|---|
| Industrial countries | 100 | 96 | 93 | 88 |
| Capital-surplus oil-exporting countries | 80 | 113 | 140 | 133 |
| Oil-exporting developing countries | 89 | 105 | 118 | 121 |
| Oil-importing developing countries | 79 | 98 | 100 | 98 |

*Source:* World Bank and International Energy Agency.
[a] Estimated.

is a definite reduction in the energy intensity of production in the industrial nations, the trend is upward for the three other groups, notwithstanding the decline between 1978 and 1980 for the capital-surplus countries and the oil-importing developing countries.

In summary, over the decades since World War II, trends in the demand for oil for different groups of countries have been diverse, bringing about varying rates of growth. The demand for energy is a function of many variables, and these should be treated in the most sectorally disaggregated form possible. Generally, the structure of demand for energy in the long run would be given by the function:

$$D = f(K, L, M, Y, P_E, P_K, P_L, P_M),$$

where $D$ is the demand for energy for any set of activities, $K$ is the stock of capital employed, $L$ is the labor employed, $M$ is the quantity of raw materials used, $Y$ is the level of output, $P_E$ is the price of energy, $P_K$ is the price of capital, $P_L$ is the wage rate for labor, and $P_M$ is the price of raw materials used.

Even though recent work using the translog form has provided very useful results from a number of research efforts,[1] most global studies use much simpler aggregation. Typically, for predicting energy demand some estimates of long-run price and income elasticities are combined with assumptions of future increases in energy prices and incomes in the region being considered. The World Bank (1981a) used this approach to develop forecasts of demand for energy by groupings of countries for the years 1985 and 1990 (presented in the following chapter).

But any forecast of the future has to be based on a very clear understanding of the past. Since price has been established in most studies as an important determinant of demand for energy, it is interesting to observe the changes in energy intensity shown in table 2.1 in conjunction with changes in energy prices. Energy prices for the years 1975, 1978, and 1980 are shown for the same groups of countries in table 2.2, treating 1973 as a base year in each case. There are substantial lags in the effects of price changes on quantity demanded, and undoubtedly the price changes between 1973 and 1975 had exhibited only part of their full long-run effects on demand by 1980. But the effects of prices are so dominant in the determination of demand that almost every major study in recent years has underestimated this effect and overshot actual values observed, particularly those occurring after 1980. Extrapolations of pre-1973 trends, which were dominated by decreases in real

---

1. See, for instance, Berndt and Wood 1976; and J. M. Griffin, "Interfuel Substitution Possibilities: A Translog Application to Intercountry Data," *International Economic Review*, October 1977, pp. 149–64.

TABLE 2.2

Index of Energy Prices to Final Users, by Country Group, 1975, 1978, 1980
(1973 = 100)

| Country Group | 1975 | 1978 | 1980[a] |
|---|---|---|---|
| Industrial countries[b] | 133 | 144 | 195 |
| Capital-surplus oil-exporting countries[a] | 90 | 80 | 70 |
| Oil-exporting developing countries[c] | 125 | 125[a] | 160 |
| Oil-importing developing countries[c] | 141 | 150[a] | 200 |

*Source*: See table 2.1.
[a] Estimated.
[b] Average of seven major industrial countries.
[c] Average based on a sample of developing countries in each group.

price of oil even when adjusted for the effects of increasing prices since 1973, were generally off the mark, as evidenced by reduced consumption since 1979.

Over the past twenty years there have also been major structural changes in the economies of most countries. In some nations, particularly in the industrialized West, these changes occurred in response to actual and anticipated changes in factor price ratios. In the developing countries, on the other hand, the economic structure has moved from a great reliance on agriculture to rapid growth in industry and other sectors, notably transport and services. While it is useful to study the history of earlier transitions from wood to coal and coal to oil, the rapidity of change in economic structures and activities makes the past twenty years particularly revealing of transitional trends in the use of energy.

## *Trends in Energy Consumption Since 1960*

Data on energy consumption and supply are published by the United Nations. Major oil companies and the World Bank compile and publish useful statistics based on the UN series. Unfortunately, reliable data are not available for the consumption of noncommercial energy such as firewood and animal and vegetable waste, and this constrains a fair assessment of the transition from noncommercial to commercial fuels in the developing countries. The size of noncommercial energy consumption in the developing world varies from around 90 percent of energy consumed in some countries of Africa to around 10 to 20 percent in middle-income countries such as Korea, Brazil, and Argentina. The two most heavily populated nations of the Third World, China and India, derive over 25 percent and 40 percent, respectively, of their total energy from noncommercial sources.[2] In fact, a recent publication places the 1980 consumption of biomass in the United States at 1.8 quads out of a total energy consumption of 78.0 quads (U.S. Department of Energy 1981, table 1-3), which amounts to 2.3 percent of the total (and is, in all probability, an underestimate). There is undoubtedly a trend toward the use of fuelwood and biomass in the residential sector of the United States, and even the industrial sector is likely to move in the same direction. Unofficial estimates indicate that in some parts of the U.S. Northwest, industrial plants are deriving as much as 25 percent of their total energy from biomass. Unfortunately, reliable analysis and estimation of substitution possibilities between commercial and noncommercial energy will have to await the development of systematic data collection and computation for noncommercial fuels.

Table 2.3 shows the consumption of energy in the industrial countries for selected years since 1960. It can be seen from these figures that the upward trend of oil consumption reversed itself after 1973. The early increase in the use of oil and its recent decline is reflected in a decrease and then a slight gain in the use of coal, while primary electricity steadily increased through the whole period. The importance of these trends to

TABLE 2.3

Energy Consumption in Industrial Countries, Selected Years, 1960–80

| Energy Source | 1960 mbd | 1960 % | 1970 mbd | 1970 % | 1973 mbd | 1973 % | 1978 mbd | 1978 % | 1980 mbd | 1980 % |
|---|---|---|---|---|---|---|---|---|---|---|
| Petroleum | 14.2 | 39.6 | 29.9 | 49.3 | 35.5 | 51.5 | 34.7 | 49.6 | 35.0 | 48.4 |
| Natural gas | 6.7 | 18.7 | 12.8 | 21.1 | 14.6 | 21.2 | 14.5 | 20.7 | 15.0 | 20.7 |
| Solid fuels | 12.3 | 34.2 | 13.3 | 22.0 | 12.8 | 18.6 | 13.2 | 18.8 | 14.0 | 19.3 |
| Primary electricity | 2.7 | 7.5 | 4.6 | 7.6 | 6.0 | 8.7 | 7.6 | 10.9 | 8.4 | 11.6 |
| Total | 35.9 | 100.0 | 60.6 | 100.0 | 68.9 | 100.0 | 70.0 | 100.0 | 72.4 | 100.0 |

*Source:* United Nations, *World Energy Supplies 1973–78*, Series J. no. 22; World Bank estimates for 1980.

2. Total primary energy consumption in China is around 860 million tons of coal equivalent per year, including about 220 million tons of noncommercial energy; based on private communications with officials of the Energy Research Association of China. For India, see "Report of the Working Group on Energy Policy" (Delhi: Controller of Publications, 1980), table 2.8.

the global energy situation is emphasized by the fact that industrial countries consume somewhat more than half the world's total energy. Apart from interfuel shifts in consumption, there has been a sharp reduction in energy intensity in the economies of these nations since 1973, as mentioned earlier.

Our study deals with countries with market economies only and excludes the centrally planned economies of the USSR and the countries of eastern Europe. But it is useful to observe some important developments in these countries, on account of their possible impacts on the energy markets of the noncommunist world. Despite the impression of a much greater control on consumption decisions, their share of oil in energy consumption continued to grow even between 1973 and 1980. Correspondingly, the share of solid fuels continued to decline in this period, and consumption of natural gas grew rapidly between 1960 and 1978, before leveling off through 1980. However, these countries retained a higher dependence on coal than did the industrial countries of the noncommunist world. In 1980, for instance, the centrally planned economies consumed only 13.1 mbd of oil, but 20.9 mbd of solid fuels, which constituted 30.5 percent and 48.6 percent, respectively, of their total energy consumption.

Table 2.4 shows trends in energy consumption in the oil-exporting developing countries. These figures show a very high relative dependence on oil and an almost negligible level of coal consumption. A number of these nations have, in recent years, followed a policy of using larger quantities of natural gas in their industrial plants, resulting in a rapid growth in consumption of this fuel, which still faces problems in transportation to export markets. The most significant feature about past energy trends in the oil-exporting developing countries is the rapid

TABLE 2.4

Energy Consumption in Oil-Exporting Developing Countries, Selected Years 1960–80

| Energy Source | 1960 mbd | 1960 % | 1970 mbd | 1970 % | 1973 mbd | 1973 % | 1978 mbd | 1978 % | 1980 mbd | 1980 % |
|---|---|---|---|---|---|---|---|---|---|---|
| Petroleum | 0.9 | 69.2 | 1.8 | 64.3 | 2.2 | 66.7 | 3.5 | 68.6 | 3.6 | 65.5 |
| Natural gas | 0.2 | 15.4 | 0.7 | 25.0 | 0.8 | 24.2 | 1.1 | 21.6 | 1.4 | 25.4 |
| Solid fuels | 0.1 | 7.7 | 0.1 | 3.6 | 0.1 | 3.0 | 0.1 | 2.0 | 0.1 | 1.8 |
| Primary electricity | 0.1 | 7.7 | 0.2 | 7.1 | 0.2 | 6.1 | 0.4 | 7.8 | 0.4 | 7.3 |
| Total | 1.3 | 100.0 | 2.8 | 100.0 | 3.3 | 100.0 | 5.1 | 100.0 | 5.5 | 100.0 |

*Source:* See table 2.3.

TABLE 2.5

Energy Consumption in Oil-Importing Developing Countries, Selected Years, 1960–80

| Energy Source | 1960 | | 1970 | | 1973 | | 1978 | | 1980 | |
|---|---|---|---|---|---|---|---|---|---|---|
| | mbd | % | mbd | % | mbd | % | mbd | % | mbd | % |
| Petroleum | 1.6 | 42.1 | 4.2 | 53.9 | 6.0 | 58.8 | 7.4 | 56.9 | 7.3 | 53.3 |
| Natural gas | 0.1 | 2.6 | 0.3 | 3.8 | 0.4 | 3.9 | 0.5 | 3.9 | 0.7 | 5.1 |
| Solid fuels | 1.8 | 47.4 | 2.4 | 30.8 | 2.6 | 25.5 | 3.3 | 25.4 | 3.7 | 27.0 |
| Primary electricity | 0.3 | 7.9 | 0.9 | 11.5 | 1.2 | 11.8 | 1.8 | 13.8 | 2.0 | 14.6 |
| Total | 3.8 | 100.0 | 7.8 | 100.0 | 10.2 | 100.0 | 13.0 | 100.0 | 13.7 | 100.0 |

*Source:* See table 2.3.

increase in consumption that took place between 1973 and 1978, amounting to a total of 54.5 percent.

The oil-importing developing countries, on the other hand, have shown a lower dependence of oil. As table 2.5 indicates, this group of countries is estimated to have registered an actual decline in oil consumption between 1978 and 1980, largely as a result of higher prices. In recent years there has also been an increase in the use of primary electricity in these nations, and large-scale investments are currently being made, which indicate that use of electricity may grow more rapidly than that of other energy sources, taking up the slack from the decline in the use of oil.

The aggregate consumption of energy continues to grow despite a downward trend in the industrial nations, which, significantly, are the world's largest consumers. The most encouraging trend since 1980 has, however, been the decrease in oil consumption, which eased the pressure on the world's oil market. Precise figures for consumption country by country are not yet available, but if oil production, which approximates consumption levels very closely, can be taken as a proxy, then oil consumption went down from 64.8 mbd in 1979, to 62.3 mbd in 1980, to 58.5 mbd in 1981, and further to 55.9 mbd in 1982. Even more significant is the decline in implied consumption of the noncommunist world from 50.57 mbd in 1979, to 47.8 mbd in 1980, to 45.0 mbd in 1981, and 41.5 mbd in 1982.

Any analysis of demand for different sources of energy is complicated by the fact that developments on the supply side cannot be easily isolated from the demand side. Consumption decisions can be influenced not only by actual shortages but also anticipated or perceived shortages. The Arab oil embargo of 1973–74 and the reduction in OPEC oil immediately after the Iranian revolution were traumatic for governments and leaders of the United States and many other countries. A whole new area of research dealing with the security aspects of energy supplies has been embarked on by many academics and strategic planners. These supply considerations have more than a minor impact on energy policies, which in turn impinge on future energy demand. The U.S. government has, with more vocal commitment than actual implementation, embarked on the establishment of a strategic petroleum reserve—in response, no doubt, to its consistent advocacy by eminent academics and oil market researchers like M. A. Adelman (1972, 215–16; 1981).

Constraints in supply of natural gas and coal, though not as dramatic as those for oil, have also influenced consumption decisions in the past. For instance, the natural gas shortages in the United States during the winters of 1976–77 and 1977–78 and coal miners' strikes in the United States, Great Britain, and Australia have produced supply interruptions or the threat of them, which adversely affected the demand for gas and coal in the long run.

## Recent Trends in Energy Supply

Among the various influences on the supply of energy, pricing is an important and often critical factor. The formation of OPEC and its emergence as the dominant force in the global oil market has been the consequence of pricing decisions in earlier periods. More recently, after the emergence of OPEC as an effective organization, prices have been closely linked with output decisions by OPEC and, more significantly, have been responsible for the expansion of oil production outside OPEC. The confusion and irrationality of natural gas regulation in the United States is the prime example of how legislative actions can interfere with an industry's free growth and result in a serious imbalance in the market through increased demand and suppressed supplies fostered by rigid controls on prices.[3]

Nuclear power is of course much too complex to be discussed in any detail in this volume. Environmental concerns in the United States, Germany, and Great Britain, in particular, have inhibited the growth of nuclear power. The nuclear proliferation problem has affected the transfer of nuclear technologies to the Third World, often stopping nuclear power development even when local governments were strongly

3. For a discussion on developments in the natural gas market in the United States and the historic dichotomy of intrastate and interstate regulation, see Stobaugh and Yergin 1979, chap. 3.

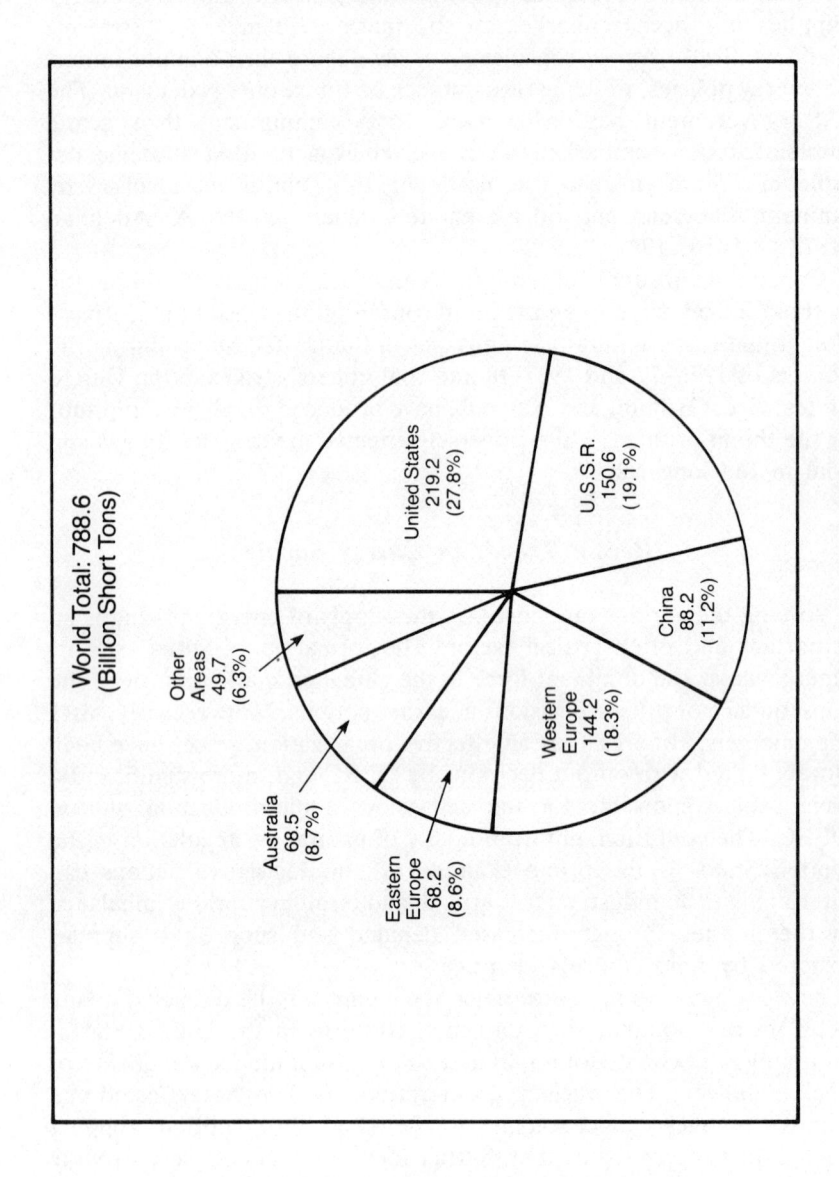

Figure 2.1 Estimated International Recoverable Reserves of Coal, 1978

committed to it and had made the required financial allocations (Ebinger 1981, 79–108). Even though in 1980, 10 percent of the world's power was from nuclear plants, less than 2 percent of the power in developing countries was produced from nuclear plants.

Over two-thirds of the recoverable coal reserves are concentrated in only four nations, the United States, the USSR, China, and Australia. A number of factors are responsible for the decline in coal production in these countries during recent decades. Environmental concerns have constrained growth, particularly in the western United States. In the United States and the USSR, oil and natural gas are more competitive. China lacks adequate transportion to distribute the coal produced in the northern provinces to other parts of the country. China also increased domestic oil reserves during the seventies. Initial growth in Australia's coal production was constrained by the small size of the country's population and the consequently small domestic demand. Figure 2.1 shows the distribution of recoverable reserves of the world's coal in 1978. Since 1974, exploration and preinvestment activities for developing additional reserves of coal have been undertaken all over the world, particularly in the developing countries. According to World Bank estimates (1981*a*, 49) an additional production of twenty-five million tons coal (equivalent to 125 million barrels oil) would be possible from these sources by the second half of the 1980s. One reason coal production has not increased significantly even in the nations that have known reserves is the long lead times in opening new mining facilities. The distribution of world coal production in 1979 is shown in figure 2.2.

Hydropower resources have remained underutilized in many countries despite the existence of many favorable sites. The development of hydroelectricity has been slowed by environmental considerations. Almost half of the hydropower potential in the world is located in the developing countries, and according to the World Bank (1981*a*, 54), only an estimated 10 percent of this potential has been developed thus far. The increase in hydropower generation between 1973 and 1979 in various countries is shown in figure 2.3. Mention must also be made of geothermal power, for which potential exists in a number of countries, including Iceland, Italy, New Zealand, the United States, Indonesia, El Salvador, the Philippines, and Turkey.

Variations in estimates of prospective supplies have been the greatest for oil and natural gas. The very nature of the business—uncertainties about future exploration efforts, the probability of new discoveries, and the geology of reservoirs from which oil is produced—make valid predictions difficult.

A significant trend in oil production is the decline in the ratio of world reserves to production. The substantial increases in reserves before 1970

22 / *The Political Economy of Global Energy*

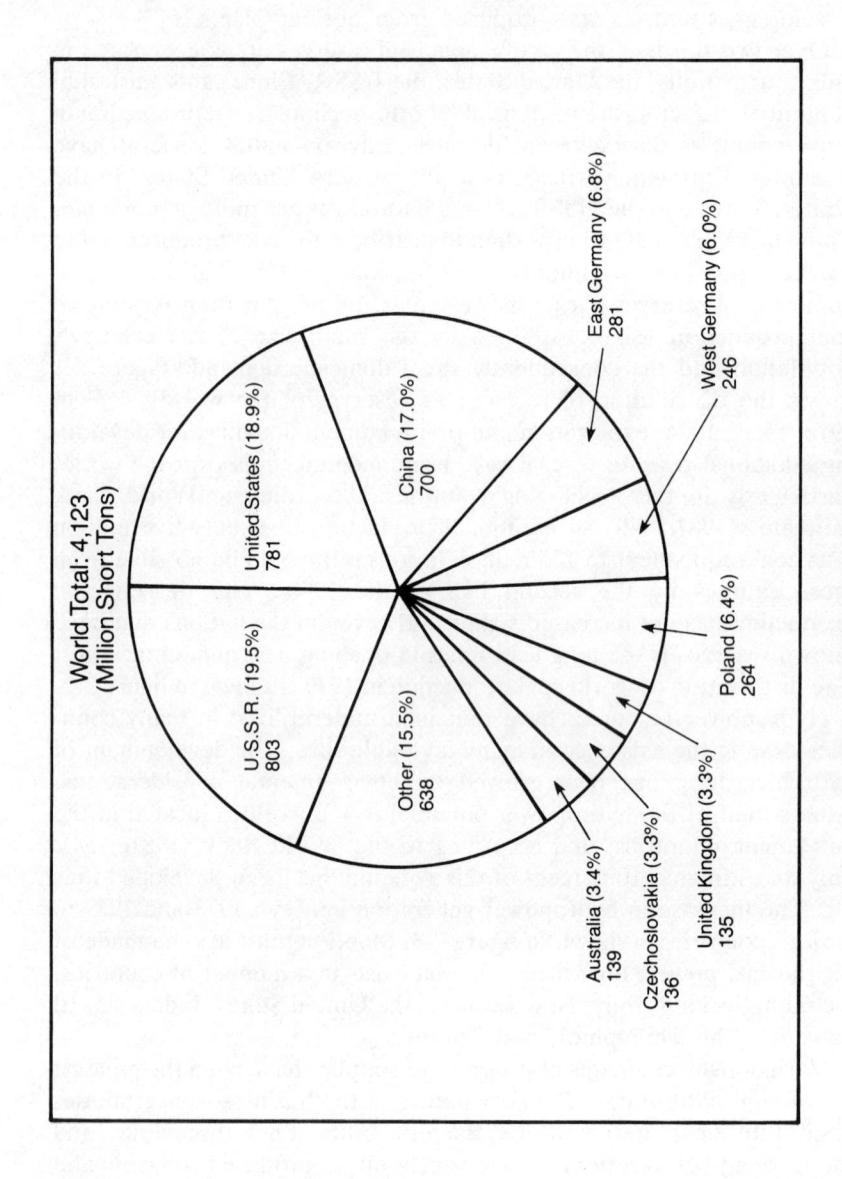

Figure 2.2 International Coal Production, 1979

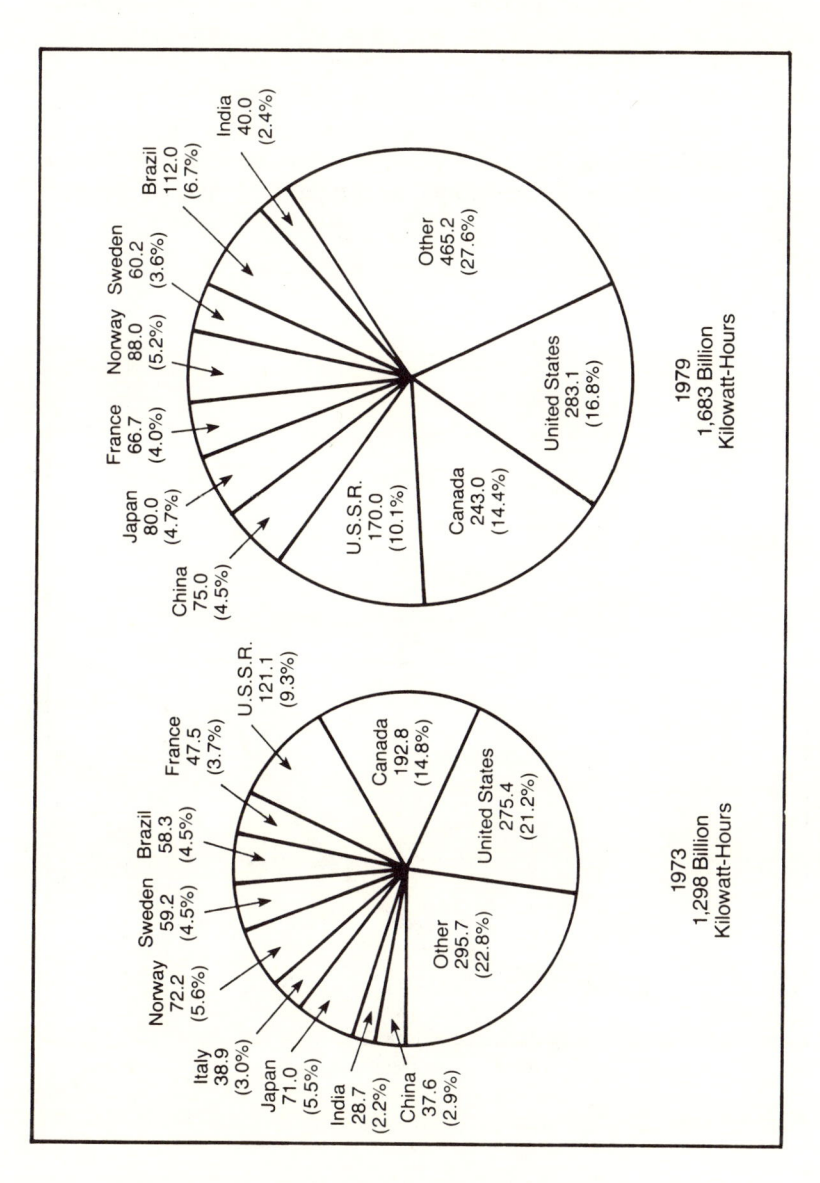

Figure 2.3 International Hydroelectric Power Production

24 / *The Political Economy of Global Energy*

can be seen in figure 2.4 to have dropped rapidly after that year. According to this chart, oil discovery rates between 1930 and 1970 were consistently in excess of production. In 1970, discoveries declined, particularly in the Middle East, and consumption substantially increased. Changes in reserves and production are shown in table 2.6. These trends and their past geographical biases reflect a pre-1973 preference by oil companies to invest in exploration in the generally more profitable Middle East, but changes in the structure of the oil industry after that have accelerated investments in nations outside

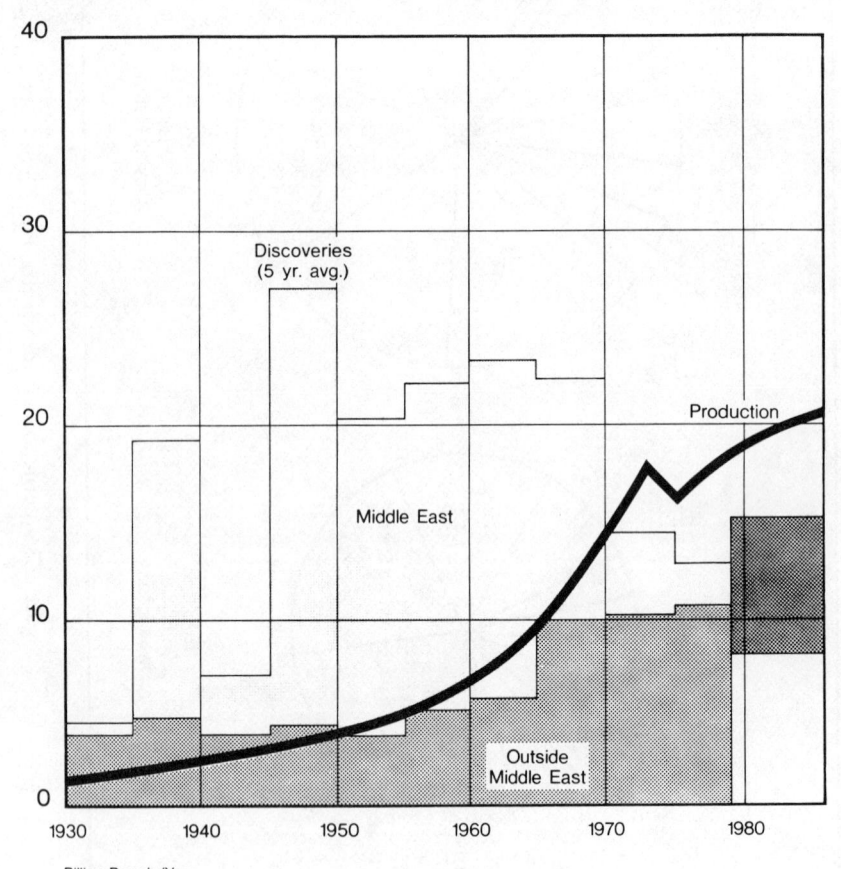

Billion Barrels/Year
*Excluding U.S.S.R. Eastern Europe and the People's Republic of China

Figure 2.4 Rate of Discovery of World Oil Reserves

*Source*: Exxon Corporation, 1979.

TABLE 2.6

Petroleum Reserves and Production, by Country Group, 1960, 1970, 1980

| Country Group | Reserves[a] 1960 | 1970 | 1980 | Production[a] 1960 | 1970 | 1980 | Ratio of Reserves to Production[b] 1960 | 1970 | 1980 |
|---|---|---|---|---|---|---|---|---|---|
| OPEC countries | 218.0 | 412.4 | 434.3 | 3.2 | 8.6 | 9.8 | 68.1 | 48.0 | 44.3 |
| Non-OPEC countries | 83.0 | 199.0 | 214.2 | 4.8 | 8.6 | 12.9 | 17.3 | 23.1 | 16.6 |
| Total | 301.0 | 611.4 | 648.5 | 8.0 | 17.2 | 22.7 | 37.6 | 35.5 | 28.6 |

*Source: World Oil*, various issues.
[a] Billions of barrels at year end.
[b] Number of years reserves would last at respective production levels.

OPEC. Table 2.7 shows the level of drilling activities in different country groups. The industrial nations and the oil-importing developing countries have apparently increased their search for oil since the early seventies. The oil-exporting nations, however, have emphasized the nonoil sectors of their economies, and their excess capacities in oil production have resulted in lower exploration efforts. The potential for new discoveries in the oil-importing developing countries remains largely unexplored despite the efforts of the last ten years. It is estimated that only 14 percent of their estimated ultimate recoverable reserves have been proven. But these figures must be regarded as

TABLE 2.7

World Drilling Effort, by Country Group, Selected years, 1965–78

| Country Group | Thousands of Feet per Year and Percentage Change[a] 1965 | 1970–72[b] | 1976–78[b] |
|---|---|---|---|
| Industrial countries | 201,169 | 153,728 (−23.5) | 254,608 (+65.6) |
| OPEC countries | 10,610 | 13,177 (+24.2) | 13,878 (+ 5.3) |
| Non-OPEC oil-exporting developing countries | 4,660 | 6,285 (+34.9) | 6,350 (+ 1.0) |
| Oil-importing developing countries | 6,928 | 7,390 (+ 6.7) | 9,904 (+34.0) |
| Total | 223,367 | 180,580 (−19.2) | 284,740 (+55.5) |

*Source*: See table 2.6.
[a] Includes exploratory, development, and service drilling; excludes the centrally planned economies.
[b] Average.

nebulous, and some discussion of the concepts involved in estimating reserves would be in order.

The term *proven reserves* essentially reflects the measure of oil that can be produced with known technology and at existing costs and prices. The oil known to exist in a specific geographical region is generally known as *oil in place*, but this oil may not all be recoverable for technological or economic reasons. The ratio between proven reserves and oil in place is generally referred to as the recovery factor, but this factor is a dynamic concept, since technical innovations in recovering oil can enhance the ratio considerably. The current status of technology in secondary and tertiary recovery can enable recovery factors in excess of 60 percent in some cases. But the precise recovery factor applicable to a specific oil field depends on very specific geological, technical, and economic features, which naturally vary considerably over regions and over time. It is, therefore, important to remember that any static assessments of proven reserves for a nation or for the world are only indications and are, therefore, a fertile subject for varying interpretations and dispute. But we will touch on these disputes while discussing the prospects of future supplies of oil.

The uneven geographical distribution of the world's oil resources and the concentration of exploration activity to limited areas has given rise to large-scale international trade in oil. The geographical imbalance in oil supply and consumption is brought out in figure 2.5, which shows that in 1978 (before the Iranian revolution) OPEC produced almost half the world's oil, while the nations in the Organisation for Economic Cooperation and Development (OECD) consumed almost two-thirds of the global output. The supply of oil in the past has, therefore, not been a direct function of market forces, and future supplies are dependent on a whole set of political and international economic issues (which are explored later in this volume). Predictions on the future behavior and levels of activity in the world oil market will be useful only if they rest on careful analysis of the major oil-producing centers of the world, OPEC and its constituents, especially.

For some years now, and particularly in response to the events since 1973, a number of oil-importing nations have accelerated efforts at promoting the use of what is commonly referred to as new and renewable forms of energy. These efforts have received varying degrees of backing from those in authority (contrast the commitment of the Carter administration with President Reagan's withdrawal of government involvement). But much of the potential of these sources of energy is in the form of theoretical availability, without any convincing results from hard economic analysis. Consequently, visions of rapid development have given way to more realistic estimates of exploitable potential,

*Global Energy Transition* / 27

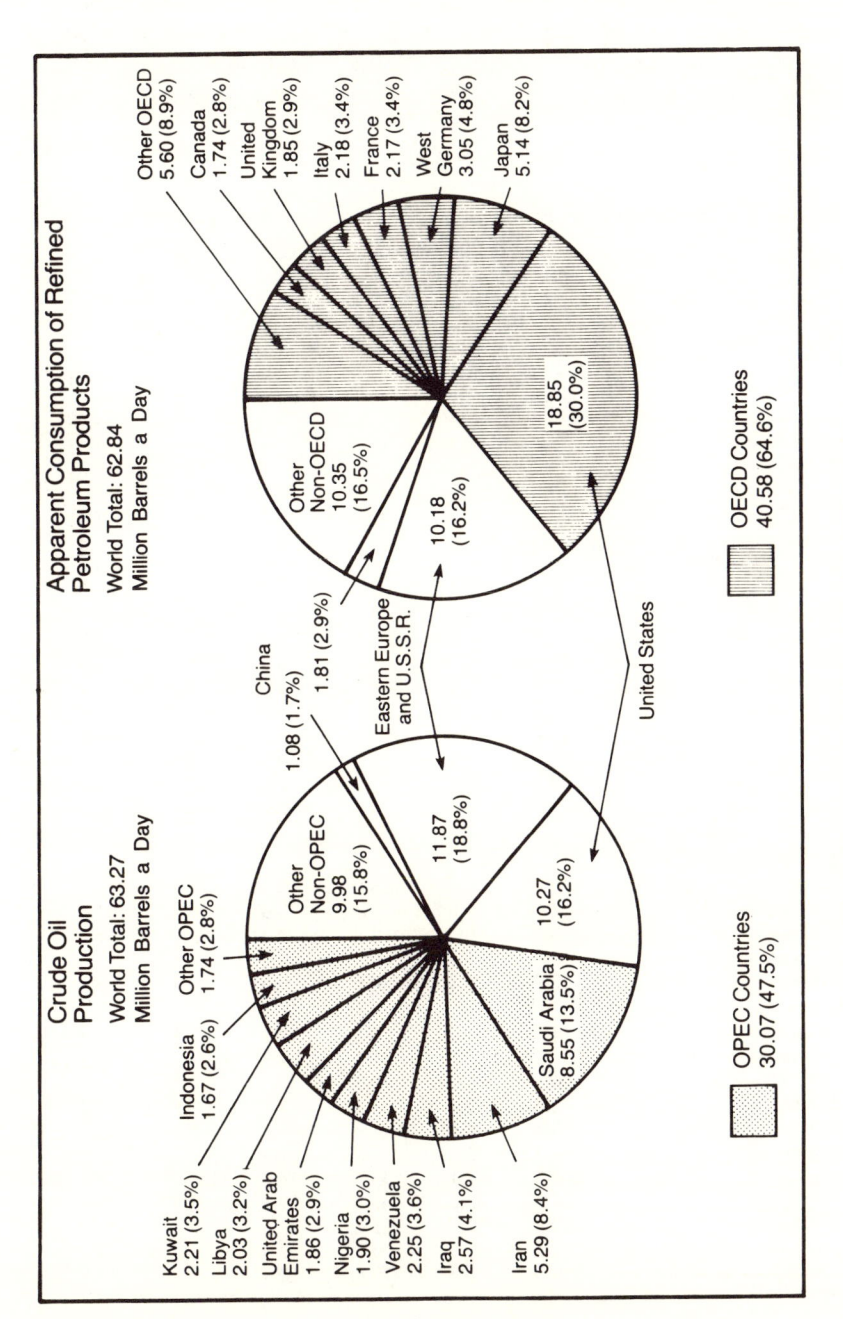

Figure 2.5 International Petroleum Supply and Disposition, 1978

28 / *The Political Economy of Global Energy*

with due regard to the enormous constraints of technology and finance standing in the way of a rapid buildup. Hence, in the medium term the role of new and renewable sources of energy is likely to be closer to "a mosquito bite on an elephant's fanny" than "forty percent of our

TABLE 2.8

Commercial Primary Energy Production and Consumption, by Country Group, 1970, 1978, 1980
(millions of barrels a day)

| Country Group | 1970 Production | 1970 Consumption | 1978 Production | 1978 Consumption | 1980 Production | 1980 Consumption |
|---|---|---|---|---|---|---|
| Industrial countries | 43.2 | 60.6 | 46.9 | 70.4 | 50.6 | 72.4 |
| Petroleum | 12.7 | 29.9 | 13.3 | 35.3 | 14.5 | 35.0 |
| Natural gas | 13.0 | 12.8 | 14.1 | 14.9 | 13.8 | 15.0 |
| Solid fuels | 13.0 | 13.3 | 12.4 | 13.1 | 13.9 | 14.0 |
| Primary electricity | 4.5 | 4.6 | 7.1 | 7.1 | 8.4 | 8.4 |
| Centrally planned economies | 28.8 | 27.6 | 44.0 | 41.7 | 45.2 | 43.0 |
| Petroleum | 8.0 | 7.2 | 14.3 | 12.7 | 13.7 | 13.1 |
| Natural gas | 3.8 | 3.8 | 7.0 | 6.9 | 7.7 | 7.0 |
| Solid fuels | 16.1 | 15.7 | 21.1 | 20.5 | 21.8 | 20.9 |
| Primary electricity | 0.9 | 0.9 | 1.6 | 1.6 | 2.0 | 2.0 |
| Capital-surplus oil-exporting countries | 12.8 | 0.3 | 17.5 | 0.7 | 18.6 | 0.9 |
| Petroleum | 12.7 | 0.2 | 17.3 | 0.5 | 18.3 | 0.7 |
| Natural gas | 0.1 | 0.1 | 0.2 | 0.2 | 0.3 | 0.2 |
| Solid fuels | — | — | — | — | — | — |
| Primary electricity | — | — | — | — | — | — |
| Oil-exporting developing countries | 13.7 | 2.8 | 18.0 | 5.0 | 16.7 | 5.5 |
| Petroleum | 12.7 | 1.8 | 15.9 | 3.3 | 14.2 | 3.6 |
| Natural gas | 0.7 | 0.7 | 1.7 | 1.2 | 2.0 | 1.4 |
| Solid fuels | 0.1 | 0.1 | 0.1 | 0.2 | 0.1 | 0.1 |
| Primary electricity | 0.2 | 0.2 | 0.3 | 0.3 | 0.4 | 0.4 |
| Oil-importing developing countries | 4.7 | 7.8 | 6.8 | 13.2 | 7.5 | 13.7 |
| Petroleum | 1.2 | 4.2 | 1.2 | 7.3 | 1.5 | 7.3 |
| Natural gas | 0.3 | 0.3 | 0.5 | 0.5 | 0.5 | 0.7 |
| Solid fuels | 2.3 | 2.4 | 3.3 | 3.6 | 3.5 | 3.7 |
| Primary electricity | 0.9 | 0.9 | 1.8 | 1.8 | 2.0 | 2.0 |
| Bunkers | — | 2.9 | — | 2.8 | — | 3.1 |
| Total | 103.2 | 99.1 | 133.2 | 131.0 | 138.6 | 135.5 |
| Petroleum | 47.3 | 43.3 | 62.0 | 59.1 | 62.2 | 59.7 |
| Natural gas | 17.9 | 17.7 | 23.6 | 23.6 | 24.3 | 24.3 |
| Solid fuels | 31.5 | 31.5 | 36.9 | 37.3 | 39.3 | 38.7 |
| Primary electricity | 6.5 | 6.6 | 10.8 | 10.9 | 12.8 | 12.8 |

*Source:* See table 2.3.

energy" (Maidique 1979, 183, on solar energy). Even though significant developments have taken place in the methods for use of solar energy and other renewable forms and some encouraging trends have been exhibited in their applications, the impact on the world's energy problem has remained small. The most promising application has been the use of solar energy for heating purposes, and the idea has been catching on visibly in the U.S. Southwest, specifically Arizona, New Mexico, Texas, and California.

In sum, trends show a remarkably swift transition in the West from fuelwood to coal in the initial stages of industrialization and, more recently, from coal to oil, bringing about an overwhelming dependence on oil. The centrally planned economies have been no exception to this trend, and even though their dependence on oil is somewhat lower, they do not show the reductions in energy intensities in evidence in the industrial nations of the noncommunist world. At the other end of the spectrum, most developing nations have been following a transition path that goes directly from the use of traditional fuels to the use of oil and natural gas. Changes in energy consumption and production between 1970 and 1980 in the various country groups is shown in table 2.8.

# CHAPTER 3
# FUTURE ENERGY PROSPECTS

Future energy prospects have been investigated in numerous studies in recent years, ranging from the Ford Foundation's energy project (Freeman 1974) in the early seventies to the recent global study of the International Institute of Applied Systems Analysis (IIASA) (Hafele 1981). The emphases and time horizons of various studies have differed considerably, but recent works have generally looked at scenarios extending to the year 2000, except in the case of the IIASA study, which looks at energy developments right through the year 2030. Undoubtedly there is much merit in futuristic studies of the IIASA mold, which focus attention on the longer term developments of importance to the future of human society. Though such studies have little direct impact on policy making and decisions on resource allocation, they do bring about a higher consciousness and put current trends in perspective, which may ultimately influence policies.

In the world of forecasting, scenarios covering anything beyond the immediate, short-term, or medium term future are increasingly regarded as hazardous territory. The global linkages influencing developments in any region or nation have grown so much in recent years that nations no longer are insulated from global events, and are, therefore, in no position to make national forecasts without allowing for global developments. In this period of rapid and often turbulent change, forecasts beyond the next six or seven years are not reliable. This book examines scenarios to the year 2000; but any prediction beyond the eighties at the present juncture is admittedly weak.

## *The Impact of Energy Policies*

Future energy scenarios are not merely a function of current trends but will be influenced by policy changes adopted by governments in the future and by structural changes in energy markets and in the consuming sectors of the world economy. The dominant influence of current trends on long-term predictions is evident from the fact that each of the recent major studies on energy future has forecast lower energy consumption in the year 2000 than the publication that preceded it. For instance, the Ford Foundation study (Freeman 1974, 13) shows U.S. energy consumption in the year 2000 to be as high as 187 quadrillion Btus; this drops to 132 quadrillion Btus in the study by the workshop on Alternative Energy Strategies (Basile 1977, 652); and to around 100 quadrillion Btus in the U.S. Department of Energy study (1981, table 1.2). To varying degrees, the scaling down of successive forecasts reflects an acceptance of reality, such as lower rates of economic growth and reductions in energy intensities.

As mentioned earlier, recent changes in energy consumption are as much the effect of higher energy prices as of the public's perception of future scarcities and possible interruptions in supply. Governments have, in general, taken an active lead in shaping policies that would influence both the demand and supply sides of energy markets in the noncommunist world. The almost complete absence of coherent energy policies before 1973 and the widespread efforts to shape and implement long-range energy policies since then makes this area the most important element in the energy picture of the future. The importance of forward-looking energy policies has been brought out by a number of opinion leaders, researchers, and institutions, among which the report of the project sponsored by the Ford Foundation and carried out by Resources for the Future deserves special attention (Landsberg 1979).

The Ford-RFF study draws on the disappointing experience of the United States in developing a suitable policy in response to the oil crisis of 1973-74. That seeming inertia had its roots in the complexities and conflicts inherent in the formulation of energy policy. For example, higher energy prices conflicted with antiinflationary goals; energy production, distribution, and consumption resulted in undesirable environmental effects; energy independence through restructuring the nation's economy and diversifying into nonconventional energy sources would cause higher costs to the economy; and investments in nuclear energy conflicted with concern for safety and growing public opposition.

There are of course critical issues of a regional or sectional nature that national energy policies cannot possibly address in full. A workshop organized by Resources for the Future and Brookings Institution (Landsberg, Dukert 1981), investigated the equity effects of higher energy prices and drew the following conclusions:

1. Differences exist between the richest and poorest in every society, irrespective of the form of government and political system in existence.

2. The worst impact from higher energy prices is on the poorest sections of society.
3. There are major differences in energy prices and supplies, leading to serious regional problems and interregional strife.[1]
4. Substitution among forms of energy is not always possible, even though there is a definite relationship between prices and supply of all forms of energy.
5. An energy ethic based on burden sharing is necessary, in acceptance of the reality that energy prices will be higher in the future.

As though the issues themselves are not complicated enough, the articulation of an effective energy policy is hampered by the lack of national institutions and mechanisms for coordinating the diverse departments that must be involved in such an effort. The case of the United States is particularly difficult on account of the separation of power between the president and Congress, not to mention the various pressure groups with enormous influence and resources. The most notable attempt to shape an energy policy in the United States was the labored and slow passage of the 1978 legislation, but this, too, fell far short of shaping coherent national energy policy. With the change in government in 1981, even these efforts have been all but abandoned, and the organizational structure established for the purpose, slothful and overgrown as it appeared to be, is on a demolition course.

The conclusions from the Ford-RFF study referred to earlier are significant, and some of them can trigger discussions relevant to this book's treatment of the subject. There is unlikely to be a real physical shortage of energy in the near future; policies will have to contend with higher costs of production and political or environmental constraints. The challenge, therefore, is to achieve a transition to higher-cost energy usage, which involves changes in the structure of energy-using activities and assets.

For the next two decades, at least, oil will remain the most important source of energy, and supplies from the Middle East will remain critically important. The recent reduced output by OPEC and rise in the share of non-OPEC production can be misleading: OPEC remains the dominant source of traded oil, and importing nations will be vulnerable to supply interruptions. The number of countries once dependent on oil

---

1. During the Arab oil embargo of 1973–74 and the lowered national speed limit of 55 mph, in the warmer Southern states of the United States automobile bumper stickers appeared with the words "Drive Sixty and Freeze a Yankee."

imports has decreased due to domestic production increases, but those nations most vulnerable in the pre-1973 period have not substantially altered their status. Instability in the countries of the Middle East or their deliberate withholding of oil remain serious threats for these oil-importing countries. In the past, serious distortions in the market have been caused by price regulations and other controls. An efficient and speedy transition can be ensured only by allowing prices to reflect scarcity, which must inevitably result in higher costs. But the equity and regional considerations related to such an action require political will and determination of a high order.

The history of energy use, particularly in the industrial nations, and the evolution of production technologies induced by low energy prices resulted in an economic structure capable of performing at higher levels of output with lower energy use. Conservation has, therefore, been identified as an all-important "source" of energy. In this direction the lags that would normally be encountered while increasing energy conservation can be shortened by effective dissemination of information and by the use of incentives. Technological progress in this century has bred a certain faith in technical solutions, which comes to the fore when society is faced with economic challenges. But technology has limitations, and in the medium term, technological solutions may not be able to bring spectacular solutions. Hence, large-scale investments in research and development may belie their initial promise if the effects of economic, political, and social constraints in their end applications are not fully appreciated. The implications of the realities mentioned above are that an effective energy policy must be based on a judicious use of market as well as government-initiated responses. But government's actions must lead to outcomes that have been clearly identified and are in society's overall interests. In the past, unfortunately, an emphasis on sectoral approaches or partial analysis has led to economy-wide effects that were clearly not intended. The regulation of the natural gas industry in the United States is a prominent, if somewhat extreme, example of government action resulting in economic inefficiency and undesirable consequences.

The constraints working on the formulation of a desirable policy package vary across societies and nations and often reflect the lack or presence of substitutes in the short run. For instance, increases in the price of kerosene in Indonesia could bring about large-scale disaffection between government and the people. An increase in electric train fares in Calcutta could send millions on a rampage of violent protest. And an increase of fifty cents a gallon in the price of gasoline, if induced by the president of the United States, would certainly usher in a new occupant in the White House at the earliest electoral opportunity. Future energy developments are, therefore, likely to be influenced by policy actions only to the extent that these actions are acceptable to society and are within the capabilities of governments to implement. In situations where institutional and political structures happen to be much too weak to manage the transition effectively, changes may occur through revolution or convulsive events. Paradoxically, in situations of deliberate and planned change, democratic structures are often more effective than authoritarian institutions, but this cannot be generalized and depends on a host of cultural and historical factors.

Prices of energy and other factors can be influenced substantially by government policies, and in the long run, pricing is by far the most effective instrument for inducing change. And even though mandatory measures and rationing can bring about quick results in the short run, they are at best emergency measures, since they may cause distortions in the long run. But pricing policy designed to provide direction to future energy supply and demand would give rise to the following questions: How are energy supply industries structured and how do they respond to price changes? How price-elastic is the demand for energy, and what are the elasticities of substitution between fuels? What are the likely constraints on future energy supply, and what pricing strategies are likely to equilibrate energy markets? What would be the effect of subsidies and taxes in achieving desirable patterns of supply and demand?

As yet there is insufficient evidence to furnish reliable answers to these questions, and most research on the subject has provided varying estimates of relevant elasticities. But these measures are at the very core of energy pricing policy, since they are supposed to reflect the magnitude, nature, and timing of price-induced changes in energy supply and demand. There has not been enough time since the 1973 reversal in energy price trends to estimate elasticity with any degree of reliability, particularly since the duration and structure of adjustment lags is not known. A study by Griffin (1979, 101–103) uses the parabolic lag structure with a cross-country sample of 1955–72 data to estimate own price elasticities for coal, gas, and oil. The influence of the length of the lag assumed is substantial, such that the own-price elasticity for coal using an immediate response (zero lag) assumption is $-0.22$, but with a four-year lag the estimated elasticity is 0.07. In some cases the results also differ significantly depending on whether a general lag structure or a parabolic form was used. Another source of doubt on most elasticity estimates based on pre-1973 data arises from the possibility that lag structures and responses applicable to an era of declining real prices may or may not apply with increasing prices.

## Future Energy Demand

While uniform estimates of energy price elasticities do not exist as yet (as they seldom do on such subjects), there is even scantier evidence on the relationship between energy and other factors of production. For instance, the energy-capital complementarity question has not been answered despite some rigorous work in the field, taking off from the Berndt and Wood (1975, 259–68) approach. In a recent study, as yet unpublished, which I undertook on behalf of RFF, India's manufacturing industries were analyzed for interfactor and interfuel substitution possibilities, using the familiar specification of translog production functions. The evidence varied substantially among the seventeeen manufacturing groups.

The reliability of elasticity estimates is of prime importance in assessing future scenarios, since the most critical question for energy policy in the future, apart from uncertain increases in supply, is that of conservation possibilities. Earlier estimates fall short of reality when one examines the downturn in energy demanded during 1981–82, but it is not clear how much of this reduction is attributable to price changes and how much is the effect of reduced economic activity and decline in income. The most balanced approach, therefore, is to use estimates that reflect a compromise among varying results.

A recent study by the Institute for Energy Analysis of the Oak Ridge Associated Universities (ORAU)[2] has developed projections of demand using the simple relation

$$E = E_o P_e^{(a+b)},$$

where $E$ is final energy demand, $E_o$ is base energy demand, $P_e$ is the 1980 price index of energy, and $(a+b)$ is the combined energy price elasticity of energy demand and aggregate output.

The demand for individual fuels is then determined through a logit equation, which is

$$e_i = \frac{C_i P_i^r}{C_j P_j^r}$$

for all $i$ where $e$ is the final share of fuels out of total energy consumption, $P_i$ is the price, $C_i$ is the base share, and $r$ is the elasticity control parameter.

---

2. "Global Energy Consumption and Production in 2000," ORAU/IEA-81-2(M), (Oak Ridge, Tenn., and Washington, D.C.: 1981).

The projections from the ORAU study using this methodology rest on critical assumptions on prices and elasticity. The aggregate energy elasticity parameter $(a+b)$ was assumed as $-0.5$ in 1985 and growing to $-0.6$ in 1990 and $-0.7$ in 2000 for both developed and developing countries. The assumption regarding the individual fuel elasticity control parameter, which essentially is a function of own-and cross-price elasticities and moderates the changes in fuel shares in response to relative price changes, is assumed to remain at $-2.0$ for all years and for both developing and developed countries.

The price assumptions driving the ORAU model include an annual price increase of 4 percent for oil, 3 percent for gas, 2 percent for coal, and 2 percent for primary electricity for the period 1980 to 2000. A second scenario assumes a more moderate growth in energy prices of 1.5 percent annually between 1980 and 1990 and 2 percent annually between 1990 and 2000. Under these conditions, the study concludes that global energy consumption will reach a level of 499.5 exajoules by the year 2000, amounting to an increase of 85 percent over the 1978 level. Thus growth in energy consumption between 1978 and 2000 would be 2.8 percent annually as compared to 4.2 percent in the preceding thirteen years. There would be significant differences among groups of countries, with gross domestic product in the OECD nations growing at 3.0 percent a year compared to 4.1 percent for the preceding fifteen years. Growth in the nonoil-developing countries of the noncommunist world is projected to slow down to 5.0 percent a year during 1977–2000, compared to 5.6 percent between 1960 and 1977. There would also be a slowing down of growth in the communist nations and members of OPEC, resulting in an overall growth rate for the global economy of 5.0 percent a year compared to the pre-1973 rate of 5.7 percent a year. These results are presented only to provide a benchmark for comparison with some of the results from other studies presented later.

A study by the World Bank (1981*a*), apparently prepared to provide energy projections for the *World Development Report, 1981*, looks at global energy prospects through 1990. The economic growth rates assumed in this study are higher than those of the ORAU study (see table 3.1).

Energy projections for 1990 and estimates for 1980 are shown in table 3.2. These projections show considerably higher rates of growth in energy demand than those of the ORAU study. But of greater significance than the aggregate growth in energy demand is the change in shares expected for the various groups of countries and for the various fuels in each country group. These projections carry some important implications. First, except for the centrally planned economies, oil will remain the dominant source of energy. More than half the

TABLE 3.1

Average Annual Growth of Gross National Product, by Country Group, 1960–90 (percent)

| Country Group | 1960–70 | 1970–80 | 1980–90 |
|---|---|---|---|
| Industrialized countries | 5.1 | 3.3 | 3.6 |
| Centrally planned economies | 4.0 | 4.8 | 3.9 |
| Capital-surplus oil-exporting countries | 10.5 | 7.2 | 5.5 |
| Oil-exporting developing countries | 6.5 | 5.2 | 6.5 |
| Oil-importing developing countries | 5.7 | 5.1 | 5.4 |

Source: World Bank 1981a.

world's consumption of petroleum takes place in the industrial nations, but by 1990 this share will drop to just less than half. Demand for oil in the future will, therefore, be heavily weighted by developments in the industrialized countries. A decline in the share of oil in all regions is accompanied by an increase in the share of solid fuels, except in the case of the capital-surplus oil-exporting nations.

Developing countries, both oil exporting and oil importing, consumed only 14 percent of the world's total in 1980, but their share will grow to almost 20 percent by 1990. This result is expected on the bases of (1) higher growth rates of gross national product than in the industrial nations, (2) growth rates of energy consumption higher than growth rates of gross national products, and (3) higher rates of urbanization, industrialization, and growth of transportation than in the past. The growth rate of energy demand in relation to historical rates is lower in regions where energy prices were more completely adjusted to world market levels than in regions where this was not the case.

One of the better studies in recent years dealing with the future of global energy is that of the U.S. Department of Energy (1981). The various growth rates projected to the year 2000 in this study are shown in table 3.3. (The study excludes the communist nations, and lumps all the countries other than those in the OECD and OPEC into a single group). The total consumption of energy by fuels is projected to grow as shown in figure 3.1. Significantly, this study predicts that oil consumption in the countries covered will remain almost constant at around fifty million barrels a day through the year 2000, although total primary energy consumption will continue to grow at the annual rates mentioned above, that is at 2 percent a year for the period 1980–2000. (See figure

TABLE 3.2
Commercial Energy Consumption, by Country Group, 1980 and 1990
(millions of barrels a day)

| Country Group | 1980 Consumption | 1980 Share of World Total (%) | 1980 Share of Total Energy for Country Group (%)[a] | 1990 Consumption | 1990 Share of World Total (%)[a] | 1990 Share of Total Energy for Country Group (%)[a] |
|---|---|---|---|---|---|---|
| Industrial countries | 72.4 | 53.4 | 100.0 | 87.0 | 47.0 | 100.0 |
| Petroleum | 35.0 | 58.6 | 48.3 | 37.4 | 51.6 | 43.1 |
| Natural gas | 15.0 | 61.7 | 20.7 | 16.2 | 47.4 | 18.6 |
| Solid fuels | 14.0 | 36.2 | 19.3 | 19.1 | 34.6 | 22.0 |
| Primary electricity | 8.4 | 65.6 | 11.6 | 14.3 | 61.6 | 16.4 |
| Centrally planned economies | 43.0 | 31.7 | 100.0 | 62.1 | 33.5 | 100.0 |
| Petroleum | 13.1 | 21.9 | 30.5 | 17.3 | 23.9 | 27.9 |
| Natural gas | 7.0 | 28.8 | 16.3 | 12.3 | 36.0 | 19.8 |
| Solid fuels | 20.9 | 54.4 | 48.6 | 29.4 | 53.3 | 47.3 |
| Primary electricity | 2.0 | 15.6 | 4.7 | 3.1 | 13.4 | 5.1 |
| Capital-surplus oil-exporting countries | 0.9 | 0.7 | 100.0 | 1.7 | 0.9 | 100.0 |
| Petroleum | 0.7 | 1.2 | 77.8 | 1.1 | 1.5 | 64.7 |
| Natural gas | 0.2 | 0.8 | 22.2 | 0.6 | 1.8 | 35.3 |
| Solid fuels | — | — | — | — | — | — |
| Primary electricity | — | — | — | — | — | — |

|   |   |   |   |   |   |
|---|---|---|---|---|---|
| Oil-exporting developing countries | 5.5 | 4.1 | 100.0 | 10.0 | 100.0 |
| Petroleum | 3.6 | 6.0 | 64.5 | 5.5 | 55.0 |
| Natural gas | 1.4 | 5.8 | 25.5 | 3.5 | 35.0 |
| Solid fuels | 0.1 | 0.3 | 1.8 | 0.3 | 3.0 |
| Primary electricity | 0.4 | 3.1 | 7.3 | 0.7 | 7.0 |
| Oil-importing developing countries | 13.7 | 10.1 | 100.0 | 24.3 | 100.0 |
| Petroleum | 7.3 | 12.2 | 53.3 | 11.2 | 46.1 |
| Natural gas | 0.7 | 2.9 | 5.1 | 1.6 | 6.6 |
| Solid fuels | 3.7 | 9.6 | 27.0 | 6.4 | 26.3 |
| Primary electricity | 2.0 | 15.6 | 14.6 | 5.1 | 21.1 |
| Bunkers | 3.1 | — | — | 4.6 | — |
| Total | 135.5 | 100 | — | 185.1 | — |
| Petroleum | 59.7 | — | 44.1 | 72.5 | 39.2 |
| Natural gas | 24.3 | — | 17.9 | 34.2 | 18.5 |
| Solid fuels | 38.7 | — | 28.6 | 55.2 | 29.8 |
| Primary electricity | 12.8 | — | 9.4 | 23.2 | 12.5 |

*Source:* See table 3.1.

[a] May not add to 100 due to rounding.

3.2.) OPEC consumption of oil is projected to grow rapidly as a result of high growth rates of economic output and subsidized energy prices. At the same time OECD consumption is projected to decline to under thirty million barrels a day by the year 2000. This is predicated on the consequences of lower economic growth, higher energy conservation, and substitution of other fuels for oil.

Some of the assumptions and conclusions of the Department of Energy study have an extreme conservationist basis. For instance, it is assumed that the OECD nations will achieve a growth ratio between energy and gross national product of 0.5, whereas this ratio was 0.95 in the period 1960–80 and around 1.1 in the pre-1973 period. Similarly, the rest of the "free-world" nations, which include the bulk of the developing countries, have been assumed to reach a value of 0.9 in the energy/GNP growth ratio for the period 1980–2000. The logic behind these assumptions is that the OECD nations have been through the phase of energy-intensive industrialization and will have the means to substitute capital for energy (which assumes that capital and energy are substitutes in the long run). On the other hand, the developing nations are entering a phase of industrialization that would not permit them to achieve results as spectacular as those of the OECD nations.

TABLE 3.3

Growth in Energy Consumption and Gross National Product, by Country Group, 1960–2000 (percent)

| Country Group | 1960–80[a] | 1980–2000[b] |
|---|---|---|
| *Primary energy consumption* | | |
| OECD countries | 3.9 | 1.2 |
| OPEC countries | 6.9 | 6.2 |
| Rest of noncommunist world | 5.8 | 3.7 |
| Average | 5.5 | 2.0 |
| *Gross national product* | | |
| OECD countries | 4.1 | 2.6 |
| OPEC countries | 5.8 | 5.8 |
| Rest of noncommunist world | 5.4 | 4.1 |
| Average | 4.7 | 3.0 |
| *Energy consumption to GNP growth ratio* | | |
| OECD countries | 0.95 | 0.5 |
| OPEC countries | 1.2 | 1.1 |
| Rest of noncommunist world | 1.1 | 0.9 |
| Average | 0.89 | 0.66 |

Source: U.S. Department of Energy 1981.
[a] Actual.
[b] Projected.

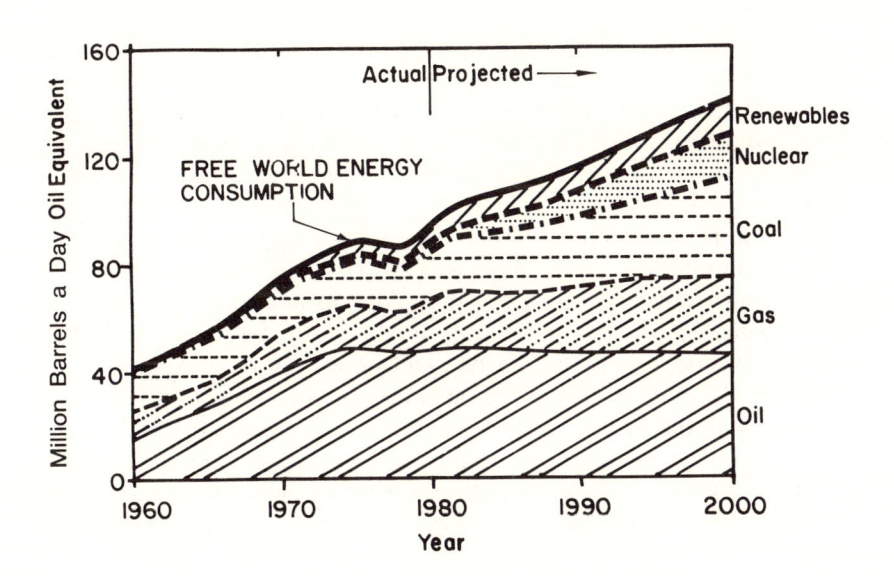

Figure 3.1 Free World Energy Consumption
*Source* : U.S. Department of Energy 1981.

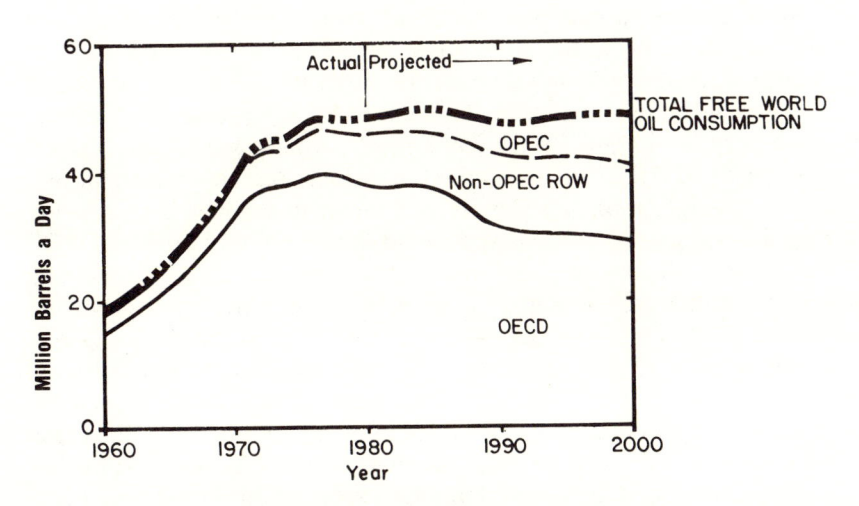

Figure 3.2 Free World Oil Consumption
*Source*: See figure 3.1.

The OECD transition has been explored by Richard Lamb (1982) who uses a reference case energy-to-gross-domestic-product ratio of 0.63 for the year 2000 for OECD nations as a whole. Lamb's scenarios, though not of unusual consequence for their predictive value, do bring out the importance of investments in various sectors of the economy to achieve a transition in energy consumption and supply in keeping with expected prices and scarcities of various fuels. He concludes that reasonably high economic growth rates would probably be required to establish an economic climate conducive to bringing nonoil energy supplies on-stream.

If economic growth is of importance in the energy transition for the OECD nations, it is an absolute necessity for the oil-importing developing countries, which not only have to import oil but have to develop indigenous energy at considerable cost, over and above the normal investments in infrastructure and industrialization. At this stage, therefore, it would be useful to look at the potential for energy supply in keeping with the scenarios of demand discussed earlier. It is only after we assess constraints and opportunities on both the demand and supply sides that we can arrive at the policy imperatives and directions to guide energy transition and the restructuring of the international economic order.

## Prospects for Energy Supply

Most studies in the past, including those undertaken during the mid-seventies, projected demand on the bases of assumed rates of economic growth, then assessed the potential for supply of nonoil energy, and finally arrived at the demand for oil as a residual source, predicated on the pre-1973 psychology of unconstrained supplies of oil (see Basile 1977). Even though this statement is perhaps an extreme and unfair generalization, oil scarcity as a factor in shaping future energy demand was grossly underplayed. Besides, the supply of oil from OPEC was seen as flowing at easily predictable levels, adequate for meeting future demand, albeit at the higher prices determined by a cartel. As late as 1978, some saw the demise of OPEC, and the declining real prices of 1977–78 brought visions of plenty to oil-market analysts. (Even as late as November 1980, Exxon Corporation was projecting OPEC output for 1985 in excess of thirty-five million barrels a day.) The cartel view is discussed in the next chapter; but whatever its label, OPEC has demonstrated that predictions for its export of oil cannot be made within the neoclassical framework of cartelized profit-maximizing behavior. A deeper, more specific hypothesis is necessary to analyze and explain the actions of OPEC, and this must embrace the political, economic, cultural, and global dimensions of the problem. Similarly, the supply of other forms of energy has to be viewed in terms of technical possibilities, institutional responses, and financial resources for development and exploitation.

TABLE 3.4
Coal Production, by Country Group, 1970, 1980, 1990

| Country Group | 1970 mbdoe | 1970 % | 1980 mbdoe | 1980 % | 1990 mbdoe | 1990 % |
|---|---|---|---|---|---|---|
| Industrial countries | 13.0 | 41.3 | 13.9 | 35.4 | 20.4 | 36.4 |
| Centrally planned economies | 16.1 | 51.1 | 21.8 | 55.4 | 29.8 | 53.1 |
| Capital-surplus Oil-exporting countries | 0.0 | 0.0 | 0.0 | 0.0 | 0.0 | 0.0 |
| Oil-exporting developing countries | 0.1 | 0.3 | 0.1 | 0.3 | 0.3 | 0.5 |
| Oil-importing developing countries | 2.3 | 7.3 | 3.5 | 8.9 | 5.6 | 10.0 |
| Total | 31.5 | 100.0 | 39.3 | 100.0 | 56.1 | 100.0 |

*Source*: United Nations, *World Energy Supplies 1973-78*, series J, no 22; World Bank projections for 1990.

The size and direction of the energy transition depends largely on the supply and use of coal. A great deal of exploration and preinvestment work has been done since 1974, particularly in the developing countries. But most of the increases in Asia and Africa are expected to come about in India and South Africa. The projections of the World Bank study for future coal production (shown in table 3.4) indicate that the share of production in the industrial countries and the oil-importing developing countries would go up slightly between 1980 and 1990, while the share of the centrally planned economies would show a slight decline. The development of coal supplies, even where large deposits are known to exist, is beset with a number of problems. Very appropriately, Stobaugh and Yergin (1979) title their chapter on coal (by Mel Horwitch) "Coal: Constrained Abundance." The accuracy of this title can be grasped from the fact that the coal industry is a vast system that produces, transports, and consumes coal. Hence, coal mining is only a small, although very important, problem in coal development.

In the United States, which has been referred to as the Persian Gulf for coal, a number of issues inhibit rapid growth of coal use. Most of the coal in the West, where coal seams lie close to the surface, can be strip-mined. However, the Surface Mining Control and Reclamation Act of 1977 requires mining companies to reclaim the land once it has been mined. The economic benefits and ease of strip mining are, therefore, offset by this requirement. A slow growth in production of western coal could restrain increased coal use in the Pacific Rim nations, some of which may have to rely exclusively on Australian coal. There are environmental factors and economic considerations also in the use and transportation of coal, which undoubtedly will dampen coal development, at least during the next two decades.

In the developing countries, the intersectoral implications of higher coal production presuppose large-scale investments in associated infrastructure, quite apart from mining facilities. Costs of coal projects range from $7,500 to $15,000 for the equivalent of a barrel of oil a day as against $3,000 to $7,000 for equivalent production from an oil project (World Bank 1981a, 51). Given the institutional problems of planning and coordinating investments in mining, transportation, and consumption of coal plus this lack of an economic incentive, these nations are unlikely to show any substantial increases in coal supply in the medium term. Significant increases in production are likely to take place only in China and India, and small increases in Brazil, Colombia, the Republic of Korea, Mexico, Romania, Turkey, Vietnam, and Yugoslavia. In the aggregate, the projections shown in table 3.4 appear optimistic, and certainly the scenario suggested by the World Coal Study (1980) that almost two-thirds of the additional energy needed in the next twenty years would come from coal appears quite unattainable, since this would require almost a hundred percent increase in coal production between 1980 and the year 2000.

Much attention has been focussed recently on the conversion of coal into gaseous or liquid fuels. But based on the current technology and projects, not more than 3 or 4 percent of the total coal produced is likely to be converted. A recent survey of production of liquids from coal, found that not only was the United States lagging behind President Carter's targets for synfuel development, but the relative stability of oil prices and worsening economic conditions have hampered work in other countries too (Stock 1982, 56–58). Large-scale commercialization of coal gasification or liquefaction would surmount many of the problems related to coal-burning, but this development is unlikely to occur without a major commitment of public funds, since the financial risks are likely to inhibit private-sector funding. But this is an issue of public policy and evaluation of security of supply, which will differ from country to country and cannot be assessed generally. The advantages of coal are enhanced if the coal is not burned directly or if its burning is a short phase in the manufacture of another fuel. Correspondingly, coal-based electric power is a much more acceptable and promising route than the decentralized burning of coal in homes and factories.

Primary electricity, made with hydro, nuclear, and geothermal power, is projected to increase significantly, but its share would decline in all but the oil-importing developing countries. Projections of primary electricity production are shown in table 3.5. Since most hydropower projects have long gestation periods, primary electricity generation

TABLE 3.5

Production of Primary Electricity, by Country Group, 1970, 1980, 1990

| Country Group | 1970 mbdoe | 1970 % | 1980 mbdoe | 1980 % | 1990 mbdoe | 1990 % |
|---|---|---|---|---|---|---|
| Industrial countries | 4.5 | 69.3 | 8.4 | 65.7 | 14.3 | 61.6 |
| Centrally planned economies | 0.9 | 13.8 | 2.0 | 15.6 | 3.1 | 13.4 |
| Capital-surplus oil-exporting countries | 0.0 | 0.0 | 0.0 | 0.0 | 0.0 | 0.0 |
| Oil-exporting developing countries | 0.2 | 3.1 | 0.4 | 3.1 | 0.7 | 3.0 |
| Oil-importing developing countries | 0.9 | 13.8 | 2.0 | 15.6 | 5.1 | 22.0 |
| Total | 6.5 | 100.0 | 12.8 | 100.0 | 23.2 | 100.0 |

*Source*: See table 3.4.

resulting from current decisions is not likely to grow rapidly until the decade 1990–2000. Besides, the capital intensity of these projects throws an enormous burden on the financial resources of developing countries. The earlier report of the World Bank (1980, 72) estimated a project cost of $ 47.45 billion for electric power development in the period 1980–85 for all the developing nations. (It is believed that these estimates have been revised upward by about 15 percent since the publication of the report.) By current indications, this level of financing may not be forthcoming in the next two or three years, and many countries may have to continue to build oil-burning and coal-burning plants, which have lower investments per kilowatt but have long-run disadvantages.

The promise of nuclear power held out in earlier decades has not been realized, and the prognosis for the medium term is for a moderate rate of growth. Major increases in nuclear power are planned only in the USSR, France, and Japan. The French policy favoring rapid nuclear growth is based on the country's vulnerability as a major importer of oil from the Middle East. This vulnerability will be largely offset by aggressive nuclear power development, which will enable France to generate over half of its power supply from nuclear sources. Indications are that nuclear capacity in the world may at best double by the year 2000 from its 1980 level. But this growth is likely to be concentrated in the very few countries that have the industrial capacity for design, manufacture, and installation of nuclear plants.

The most problematic form of energy in the medium term, and the one on which the global energy supply hinges, is hydrocarbon fuels. The uncertainties of oil gas supplies in the period to the year 2000 arise out of the fact that production and international trade are dependent on policies and events that defy analysis and prediction. The question of recoverable reserves does not appear to be of great direct relevance in the medium term, since there is adequate excess production capacity in existence, but the level of current reserves and the rate of new discoveries do influence producer behavior very strongly. The crucial question, How much oil does the world have? crops up in discussions of future supply, and reflects concerns about the rate of new discoveries on which, naturally, production rates will depend.

While most analysts agree that oil reserves are unlikely to be added to current stock at rates witnessed during the sixties, there are some who believe that the oil scarcity syndrome gripping the world has no basis. Peter Odell (1970) has been at the forefront of this view for many years now and his arguments merit some attention. Odell's thesis rests on the view that there is a bias in data put out by the oil companies, which essentially reflects their perception of self-interest. Prior to 1973, the oil companies discounted those regions in the world as potentially oil bearing where, for political or other reasons, they had been denied access. In recent writings by Peter Odell (1970; and Rosing 1980; in Tempest 1981), he suggests that the global scarcity of oil is a myth perpetrated by the oil companies, since they have been displaced by OPEC as major actors in the oil business. In order to create investment opportunities in other sources of energy, they have found it to be in their interests to convince the world that oil is running out and that other energy sources need to be developed urgently.

Odell's assessment, based on the simulations carried out by him, is that there is a 90 percent probability of oil production continuing to grow to the year 2009 before peaking and a 10 percent probability that the peak will be reached as late as 2077. Odell's contention is that the oil companies, which are the only effective institutions for oil exploration in the noncommunist world, are reluctant to undertake a large-scale global effort since they see no benefits to themselves.

In consonance with Odell's views, some small independent oil companies have emphasized for some time now that the potential undiscovered reserves in the world remain virtually unexplored. Foremost among those who advance this belief is Michel Halbouty, who has come to wield considerable influence in the Reagan White House and who is known to have had a major hand in the U.S. stand against establishing an energy affiliate in the World Bank for promoting energy developments in the Third World.[3] This view, if valid, would have important implications for the future of world oil, because, given the

3. See "Energy Problems of the Third World," *Petroleum Economist* 53(1), 418–19.

necessary investments and efforts, non-OPEC countries might increase their oil production substantially. This school of thought holds that huge discoveries are likely in the Gulf of Mexico, Southeast Asia, and the East China Sea, and smaller discoveries in various other parts of the world outside the Middle East.

A somewhat different view on oil resources is available in Richard Nehring's (1978) work, which discounts the probability of discovering giant oilfields. According to the author, the recoverable reserves in the world are 1,011.5 billion barrels (of which over 480 billion barrels will have been produced by the end of 1982). In his estimate, the potential for new discoveries is in the range of 688 to 1,288 billion barrels. Nehring contends that most discoveries will be small oilfields, and discoveries of giant and super-giant oilfields, that is, those of over 0.5 billion and 5.0 billion barrels capacity, respectively, will diminish substantially. Since 75 percent of the world's reserves exist in giant and super-giant oilfields, reduction in their rate of discovery would slow down the accretion to current reserves appreciably. Besides, Nehring believes that the probability of finding oilfields of the giant and super-giant class is greatest in the Middle East, particularly around the Persian Gulf.

The prevalent view of large oil companies is reflected in the following statement:

> The world's remaining conventional oil resources are assessed to be in the range of 1 to 1½ trillion barrels. This number includes oil which has yet to be discovered. Many of the best exploration prospects are situated in remote locations or harsh operating environments, as in the Arctic, where finding and developing oil will be difficult and costly. Moreover, in those areas where production already exists, future discoveries are anticipated to be smaller, on average, than past discoveries (Exxon 1979, 22).

Exxon concludes that even with a very active exploration effort, the average discovery rate to the year 2000 is likely to be well below the expected production rate of about twenty billion barrels a year. Their prediction follows, therefore, that a decline in production will start around the turn of the century.

This outlook, whether accepted or not, raises questions about geographical patterns of future oil production. If major oil-producing areas continue to be distinctly different from major consuming areas, as at present, then the ratios between production and reserves will reveal nothing about the actual periods or magnitudes of decline in oil production in the future. It would, therefore, be useful to examine the geographical distribution of the world's known reserves. These are shown in table 3.6. It is clear that if reliance is to continue to the year

TABLE 3.6

Known Recoverable Oil Reserves, by Country Group, End of 1981

| Country Group | Billions of Barrels |
|---|---|
| OPEC Persian Gulf | 358.0 |
| Iran | 57.0 |
| Iraq | 29.7 |
| Kuwait | 64.5 |
| Qatar | 3.4 |
| Saudi Arabia | 164.6 |
| United Arab Emirates | 32.3 |
| Neutral Zone | 6.5 |
| Other OPEC | 78.7 |
| Algeria | 8.1 |
| Ecuador | 0.9 |
| Gabon | 0.5 |
| Indonesia | 9.8 |
| Libya | 22.6 |
| Nigeria | 16.5 |
| Venezuela | 20.3 |
| Total OPEC | 436.7 |
| Non-OPEC countries | 233.0 |
| Total | 669.7 |

Source: *Oil and Natural Gas Journal*, December 28, 1981, pp. 86–87.

2000 on known reserves of oil, then OPEC would remain in a dominant position. Even should non-OPEC nations produce at rates higher than OPEC's, it would be difficult to sustain these due to technical constraints. Within OPEC, the OPEC Persian Gulf countries hold 82 percent of total known reserves, which underlines the crucial position of these nations in any future assessment of oil production.

My assessment of probable reserves is that reality lies somewhere in between the oil industry view and the Odell scenario. But geologists, geostatisticians, economists, and politicians will continue the dispute for years to come, and precise numbers are neither likely to be agreed on nor of immediate relevance. What is of much greater relevance, however, is the fact that the term *constrained abundance* used in connection with coal is particularly applicable to oil when assessing possible production to the year 2000. These abundances and constraints consist of (1) considerable excess capacity among the members of OPEC, but a growing determination among them to limit production for political and economic reasons; (2) a sizable potential for discoveries and higher production outside of OPEC, particularly in the developing countries, but a serious lack of financial and institutional capacity for exploration and exploitation of this potential; (3) an abandonment of development of shale oil and synfuels, arising out of the U.S. government's withdrawal from this field and the reluctance of the private sector to pursue it alone.

Given the likelihood that present structures and institutions will remain unchanged over the next two decades, there appears little hope of any sharp departure from the present path. The world's dependence on existing sources of supply will probably continue undiminished. Such a prospect would leave the OPEC Persian Gulf nations in a position of unprecedented strength for determining prices and supply. It appears unduly optimistic, therefore, to believe that oil supplies from OPEC will be in excess of twenty million barrels a day by the year 2000 (to balance demand and supply). Yet even the most pessimistic of projections in recent years have assumed this output from OPEC as the floor on production in the year 2000, with corresponding predictions placing the share of OPEC Persian Gulf nations at around two-thirds of this output. Some well-known projections holding on to the twenty-million-barrel-a-day floor are those by the U.S. Department of Energy (1981), the Exxon Corporation (1981), and Nordin Ait-Laoussine.[4] To conform to these forecasts, I have developed a scenario for OPEC's output of oil to the end of this century (see table 3.7). The basis for this blinkers-on scenario is a disregard for the global implications and internal socioeconomic impacts on members of OPEC of these output levels. I later examine the scenario for its effects on the international economic order and associated monetary flows.

In this discussion on supply we have not thus far touched on the potential for natural gas production. World natural gas reserves are currently estimated at around 2,600 trillion cubic feet. The significance of these reserves is that they equal 72 percent of proven oil reserves and 15 percent of proven coal reserves in the world. The distribution of these reserves, as shown in table 3.8, is such that over 75 percent of them are located in North America, the Middle East, and the centrally planned economies. The prognosis for discovery of new reserves is bright, and it is likely that the world's ultimate recoverable reserves will turn out to be around four times the current level. Unfortunately, since natural gas is a byproduct of oil, a significant portion is lost through flaring. Most of this gas in the industrial nations is utilized, but the lack of domestic markets and the high costs of liquefaction and transportation result in other regions of the world flaring most of their output. This is likely to change with downstream investments in gas-based industries by producing countries, particularly in the Middle East, and with the growing international trade in natural gas.

4. Nordin Ait-Laoussine, quoted in "Nigeria and North Sea settled at $36.50 per barrel," *Middle East Economic Survey* 25(5).

TABLE 3.7

Blinkers-on Scenario, OPEC Oil Production, Selected Years, 1980–2000
(millions of barrels a day)

| Country Group | Output of Oil[a] | | | |
|---|---|---|---|---|
| | 1980 | 1985 | 1990 | 2000 |
| OPEC Persian Gulf | 19.8 | 19.1 | 19.3 | 19.3 |
| Iran | 1.6 | 2.5 | 3.0 | 3.0 |
| Iraq | 2.6 | 3.5 | 3.5 | 3.5 |
| Kuwait | 1.4 | 1.0 | 1.0 | 1.2 |
| Libya | 1.8 | 1.2 | 1.0 | 1.0 |
| Qatar | 0.5 | 0.5 | 0.4 | 0.3 |
| Saudi Arabia | 9.7 | 8.5 | 8.5 | 8.5 |
| UAE | 1.7 | 1.4 | 1.4 | 1.4 |
| Neutral Zone | 0.5 | 0.5 | 0.5 | 0.4 |
| Other OPEC | 7.3 | 6.6 | 6.1 | 4.7 |
| Algeria | 1.0 | 0.8 | 0.8 | 0.5 |
| Ecuador | 0.2 | 0.2 | 0.2 | 0.1 |
| Gabon | 0.2 | 0.2 | 0.2 | 0.1 |
| Indonesia | 1.6 | 1.5 | 1.3 | 1.0 |
| Nigeria | 2.1 | 1.8 | 1.6 | 1.4 |
| Venezuela | 2.2 | 2.1 | 2.0 | 1.6 |
| Total OPEC | 27.1 | 25.7 | 25.4 | 24.0 |

Source: U.S. Department of Energy 1981, table 2-5, for 1980 figures. Author's compilations for 1985, 1990, 2000.
[a] Figures do not include natural gas liquids, which amounted to 0.8 mbd in 1980 and are projected to increase to 1.4 mbd in 1985, 1.8 mbd in 1990, and 2.2 mbd in 2000. 1980 figures are actual.

TABLE 3.8

Natural Gas Reserves, by Country Group, 1970 and 1980 (trillion cubic feet)

| Country Group | 1970[a] | 1980[a] |
|---|---|---|
| Industrial countries | 491.63 | 465.22 |
| Centrally planned economies | 440.00 | 953.90 |
| Capital-surplus oil-exporting countries | 162.25 | 277.63 |
| Oil-exporting developing countries | 445.25 | 856.22 |
| Oil-importing developing countries | 49.26 | 85.53 |
| Total | 1,588.39 | 2,638.50 |

Source: "Worldwide Issues," *Oil and Gas Journal* 78 (52).
[a] End of year.

Based on the blinkers-on projections, I put together the aggregate production of primary energy to the year 2000. The projections for energy demand in 1990 shown in table 3.2 forecast a demand of 185.1 million barrels a day of primary energy for the whole world. World

Bank figures show an excess of production over consumption, with the former being forecast at 189.7 million barrels a day of primary energy. These demand projections are based on fairly realistic assumptions about economic growth rates, but I visualize a much lower level of energy supplies in the future. The constraints on the oil market, on coal consumption and production, and on primary electricity resources are unlikely to permit a buildup in aggregate supply as projected. My assessment favors a supply scenario somewhat closer to the U.S. Department of Energy forecasts. These projections are shown in table 3.9.

TABLE 3.9

Projections of Primary Energy Production, by Country Group, Selected Years, 1980–2000

(millions of barrels a day)

| Fuel and Country Group | 1980 | 1985 | 1990 | 2000 |
|---|---|---|---|---|
| Oil | 49.0 | 49.5 | 50.0 | 49.0 |
| OECD countries | 15.5 | 14.5 | 14.6 | 14.8 |
| OPEC countries | 27.8 | 27.1 | 27.2 | 26.2 |
| Rest of noncommunist world | 5.7 | 7.9 | 8.2 | 8.0 |
| Coal | 17.4 | 22.2 | 25.9 | 36.8 |
| OECD countries | 14.9 | 18.7 | 20.5 | 27.8 |
| OPEC countries | — | — | 0.5 | 1.2 |
| Rest of noncommunist world | 2.5 | 3.5 | 4.9 | 7.8 |
| Gas | 16.6 | 20.6 | 24.3 | 27.8 |
| OECD countries | 14.4 | 13.8 | 17.1 | 14.7 |
| OPEC countries | 1.1 | 4.2 | 4.0 | 8.6 |
| Rest of noncommunist world | 1.1 | 2.6 | 3.2 | 4.5 |
| Nuclear | 3.0 | 5.7 | 9.4 | 15.8 |
| OECD countries | 2.9 | 5.5 | 8.9 | 14.4 |
| OPEC countries | — | — | — | 0.3 |
| Rest of noncommunist world | 0.1 | 0.2 | 0.5 | 1.1 |
| Renewables/Other | 7.0 | 7.8 | 9.0 | 14.1 |
| OECD countries | 5.4 | 6.2 | 6.9 | 10.6 |
| OPEC countries | 0.1 | 0.2 | 0.2 | 0.3 |
| Rest of noncommunist world | 1.5 | 1.4 | 1.9 | 3.2 |
| Total | 93.0 | 105.8 | 118.6 | 143.5 |
| OECD countries | 53.1 | 58.7 | 68.0 | 82.3 |
| OPEC countries | 29.0 | 31.5 | 31.9 | 36.6 |
| Rest of noncommunist world | 10.9 | 15.6 | 18.7 | 24.6 |

There are five general implications of this imbalance (an imbalance, in fact, likely to be larger unless corrective policy measures are taken early).

1. Energy scarcities will show up in energy markets from time to time and will result in physical shortages as well as rapid price increases between now and the year 2000.
2. Economic growth rates will be lower than those forecast by the World Bank in relation to the projections shown in table 3.2.
3. Because of higher oil prices and the slump in oil revenues during 1981–83, OPEC nations will reduce subsidies on oil products for domestic consumption. This will lead to lower rates of growth in OPEC's domestic consumption of oil.
4. The oil-importing developing countries will be most severely affected by energy shortages, delaying the transition from traditional fuels to commercial energy and increasing the pressure on forests and other firewood sources. Economic development will be adversely affected, and the possibility of internal strife and external interventions considerably enhanced.
5. There will be a growing incentive for the nations of the noncommunist world to import energy from centrally planned economies, and bilateral deals involving long-term arrangements are likely to increase. (The Soviet natural gas pipeline is an example.)

Without any intent to paint a picture of doom or cause undue alarm, I only suggest on the basis of the world energy outlook that the world must be aware of the grim prospects ahead. It is only from such a realization that organizations and leaders will forge institutional solutions that may avert crises situations. Some of these solutions are presented in later chapters.

CHAPTER 4

# THE HISTORY AND EVOLUTION OF OPEC

In the history of the world's oil industry, the emergence of OPEC is without parallel in its impact on the global economy and the uncertainty it has introduced into the future of energy supply. An understanding of the evolution and structure of OPEC is, therefore, critical to analyses of future options and policies in the energy field and of international economic relations. And an understanding of OPEC would be inadequate without discussing the history of the oil industry before the OPEC era, for in that period lie many explanations for OPEC's behavior and actions.

### *Historical Trends in the Oil Industry*

Of the many successful business enterprises in the twentieth century, the oil industry appears unique in its phenomenal growth, its extensive geographical spread, and the power it wields far beyond its own domain of operations and direct influence. It would be simplistic to explain the dominant role of the oil majors in business, government, and economics merely as a function of their market power; in fact, the oil majors' historic role results from a variety of factors, not the least of which is the politics of the regions in which they operate. Many splendid treatises document the history of the oil industry (Adelman 1972, N. H. Jacoby 1974, Mikdashi 1972, Rouhani 1971, Sampson 1975). But it is sufficient to recount briefly some salient features and events of special relevance to this study.

In the first part of this century the oil industry was highly concentrated, with strong barriers to entry. In the fifties and sixties, this structure gradually loosened, perhaps an inevitable result of the rapid growth of oil consumption and supply during this period (Adelman 1972). There is little evidence of economies of scale in the oil industry (in fact, the concept of economies of scale hardly lends itself to the oil industry, because of dispersed operations, technological differences, and the variable nature of oil reserves). The high levels of concentration can only be explained by the technological strengths of those companies. These strengths were the original basis for the horizontal growth of the oil industry, particularly in the Middle East. But in this technical specialization also lies the reason for the adversary relationship that developed between the major oil companies based in the industrialized West and governments of producer countries of the Third World. These governments found it difficult to accept not only the unfavorable economic impact of oil operations (conducted in their territories by companies with extraterritorial political loyalties) but also the exclusion from the know-how for managing their own oil-producing assets. In retrospect, the oil companies were guilty of serious insensitivity to local sentiment and aspirations, which ultimately resulted in a severe setback for not only these companies and their long-range plans but the entire world, which had sunk into a dependence on oil nurtured by the expectation of uninterrupted, low-cost supplies.

The oil industry has in the past been dominated by the seven international majors (often referred to as the seven sisters), five of which, namely, Exxon, Mobil, Gulf, Texaco, and Standard Oil of California, are based in the United States; the other two are British Petroleum, of the United Kingdom, and Royal Dutch Petroleum Company and Shell Transport and Trading, based in the Netherlands. The history of the seven sisters has been described in dramatic but revealing detail by A. Sampson (1975, ix), who labels his narrative "one of the oddest stories in contemporary history: of how the world's biggest and most critical industry came to be dominated by seven giant companies." What ultimately led to convulsive change in the global oil market and influenced the energy policies of nations round the world was the growth of their oil production in the Middle East.

The earliest phase of the international oil business was marked by the struggles of the large firms to extend their control to markets outside the United States and Europe. Price wars were followed by market agreements covering India, Southeast Asia, and the Pacific Rim nations in the Far East; the major competitors were Standard Oil, Royal Dutch/Shell, and Burmah Oil. Often these alliances led to the formation of combines, such as Burmah-Shell, which operated from 1928 until very recently.

The marketing wars were soon overshadowed by changes in the production end of these companies' activities and the increase in oil produced in the Middle East. The first major oil field in the region was discovered in Persia in 1908 by W. K. D'Arcy. The Anglo-Persian Oil Company was then formed, partly financed by Burmah Oil. During this period the British Navy converted to oil from coal (Penrose 1968, 57), and the British government bought a controlling in interest in Anglo-Persian Oil. In fact, by virtue of their colonial powers over many governments in the Middle East, British and other European interests had a clear edge over American firms in vying for control of oil resources in the region. But the importance of oil became apparent to all nations in World War I. According to Lord Curzon, "The Allies floated to victory on a wave of oil" (N. H. Jacoby 1974, 27). And in the twenties and thirties, a strong rivalry was set in motion for access and control of the oil-rich nations of the Middle East. Government support of companies was substantial, and helped U.S. companies overcome their disadvantage and get into Iraq, Kuwait, and Saudi Arabia.

Concurrently, U.S. companies also managed to gain control of most Venezuelan oil, and by World War II were therefore in a position comparable to producers of oil from outside North America and Europe. According to Edith Penrose (1968, 59), in 1938 Venezuela was the largest oil producer outside the United States and the USSR, with 515 thousand barrels a day, or one-third of total oil produced outside these two nations. Iran and Indonesia followed with 210 thousand and 150 thousand barrels a day, respectively. The United States was still the dominant producer of oil in that year, with a total output of 3.5 million barrels a day out of a world total of 5.6 million barrels a day, while the USSR came next with 570 thousand barrels a day. Other producers were relatively small, with Iraq producing 90 thousand barrels a day and Bahrain with 20 thousand barrels a day. Saudi Arabia was still a beginner in the game, and Kuwait had not yet entered the comity of oil-producing nations.

Oil industry developments during World War II were disrupted in Indonesia and Burma due to Japanese occupation, while transportation difficulties hampered supplies from the Middle East. But soon after the war, the discovery of vast oil fields in Saudi Arabia, Kuwait, and other Persian Gulf countries brought the prospects of enormous production increases at relatively low marginal costs. The marketing skills of the international majors and the accommodation and compromise among them resulted in integrated marketing and production, with a large share of development directed to oil fields in the Middle East. With an impending decline in production in the United States, the exploitation of oil provinces outside, at significantly lower costs, was attractive.

Costs and prices in the oil industry are fully treated in chapter 5. Here only their influence on the evolution of OPEC need be considered. First, I want to emphasize that oil production is not a natural monopoly as is, say, an electric power industry, with its clearly high proportion of fixed to variable costs (Adelman 1972, 45). Although the data on costs is "grievously imperfect," Adelman analyzes operating and development costs based on information available. The evidence, according to him, shows that in the post-World-War-II phase, the acceptable rate of return for development of oilfields outside the United States was lower than those in the United States, largely because geological and technical risks were lower overseas. With oil production outside the United States increasing its share in world output, the revenues grew correspondingly and provided a basis for conflict between the oil companies and host countries. The conflict centered around the pricing of crude oil in the international market. The final round was the emergence of OPEC in 1960 and its ultimate control of pricing.

The system of pricing followed by the oil majors had elements of irrationality and a preponderance of self-interest. To start with, there were essentially three oil prices—the market price, the transfer price, and, most important, the posted price. The market price, as the term signifies, was the price of crude oil under free-market conditions.[1] In the present context, the price of oil in the spot market approximates most closely the concept of market price. The bulk of oil traded internationally today, however, is governed by contracts, which, though no doubt influenced by the spot market, are often at considerable variance with it. Transfer prices were essentially book value, which permitted an oil company to obtain the maximum tax advantage. This was achieved by distributing its overhead and head office costs over its worldwide operations to maximize after-tax profits for the parent company as a whole.

The most important pricing practice was the use of the posted price. The evolution of this concept begins with the earliest operations of the international oil majors outside the United States and Europe. These companies first entered the Middle East through concessions allowing them to develop the region's oil. The host governments were compensated by payments of royalty, which was a fixed amount per barrel of oil produced. In the period preceding World War II, royalties were generally twenty to twenty-five cents a barrel (Park 1976, 12). The implications of a fixed royalty in an era of increasing prices were that producing countries suffered a worsening of their terms of trade when, paradoxically, oil prices were increasing in the international market. The royalty arrangement was widely resented by oil-exporting countries, but they lacked the will and strength to change it; furthermore, oil companies wielded considerable power in the decision-making structures of many Middle East countries. The first shot in the price

---

1. These conditions applied to a very small, arms-length, free market only; sales were to third parties or between oil companies on a net back basis.

revolution was fired by the government of Venezuela through the imposition of a tax on income earned by oil companies operating in that country. This tax was gradually increased, resulting in payments of seventy-five cents a barrel to the government. At the same time, Venezuela consolidated its hold on oil production by granting short-term licenses for developing specific regions, as opposed to the past practice of blanket long-term concessions that allowed the oil companies considerable latitude in determining the manner and extent of development. The importance of Venezuela's action was its impact on the rest of the oil-producing countries outside the United States. Venezuela vigorously promoted the fifty-fifty concept, which allowed host governments to earn 50 percent of total production profits in the form of taxes. In 1949, Venezuela sent a mission to the oil-producing countries of the Middle East with the good word about its own experience with this arrangement. Very soon thereafter, most oil countries in that part of world changed to and enforced the fifty-fifty regime (see Fesharaki 1976).

While these developments were taking place in the international arena, the role of the oil majors was diminishing for a variety of reasons. In the United States they were faced with antitrust actions plus the growth of smaller companies (often referred to as the independents). Furthermore, they were preoccupied with overseas operations, leaving regions in the United States open to the independents. In addition, European-based companies, supported as they were by their own governments (for example, France and Italy), provided greater competition. Also, U.S. independents extended their operations overseas, and were welcomed by governments that had found the majors intractable in their stands and menacing in their power.

These factors brought about a qualitative and psychological change in the dealings between oil companies and a number of oil-producing countries. In addition, the change in profit sharing made posted prices of great importance for host governments, since this determined their revenues. A commonality of interest among oil-producing countries grew quickly, increasing communications and a desire to coordinate oil policies. The Arab nations in particular saw a potential for greater control of the oil market, and in 1959 the Arab League convened the first Arab Petroleum Congress (to which some non-Arab oil-exporting nations were invited as observers). This meeting unanimously agreed to the formation of an organization of oil-producing countries, but its establishment might have been delayed in procedural tangles and the inertia of negotiations had it not been for a unilateral step taken by the oil companies in the summer of 1960. This was a time when, due to a variety of factors, including U.S. import policies, a glut was developing in the oil market, with consumption lagging significantly behind production. The oil companies cut the posted price approximately 6 percent without discussing it with the producer countries, which stood to lose large amounts of revenues. The oil-producing countries acted swiftly, and on September 14, 1960, representatives of Iran, Iraq, Venezuela, Saudi Arabia, and Kuwait, meeting in Baghdad, agreed on the establishment of the Organisation of Petroleum Exporting Countries (OPEC).

The mere establishment of OPEC, important as it was in symbolic terms, did little to materially alter the power balance between the oil companies and host governments. But it did have enough power to prevent further reductions in posted prices; thus, in a decade of declining market prices, market shares for the Arab nations increased. These countries saw their vast oil reserves as a way to make their voices heard in the politics of the region. This realization reached a peak in the 1967 Arab-Israeli war, and in 1968 Saudi Arabia, Kuwait, and Libya launched the Organisation of Arab Petroleum Exporting Countries (OAPEC). Both OPEC and OAPEC have since grown in membership, and by virtue of the commonality of some of their objectives and membership, increased sizes have not resulted in any serious conflicts between the two groups.

OPEC's posture, largely defensive in the sixties, underwent a radical change at the beginning of the seventies. In 1970 the Libyan government under a new revolutionary leader, Muammar Qaddafi, initiated negotiations with the oil companies for an upward revision in prices. The Libyans injected an element of belligerence in their negotiations, in keeping with the now well-known style of their leader; they threatened to cut production, claiming that production rates were too high for efficient recovery of reserves. Their priority target in the initial stages was Occidental Petroleum, which had reached a peak production level of 797 thousand barrels a day in April 1970. Libya forced the company to cut back production to 360 thousand barrels a day in August 1970 (Mikdashi 1972, 147). These developments took place against a backdrop of great European dependence on Libyan oil, which, with the closure of the Suez Canal in the aftermath of the 1967 Arab-Israeli war, was easier and cheaper than Saudi Arabian oil to transport. Additionally, Libyan oil has a low sulphur content, which refineries in Europe were geared for.

The Libyan-Occidental price skirmish was resolved as a result of an accident in May 1970. The pipeline that transported Saudi Arabian oil to the Mediterranean was damaged in Syria. The Syrian government did not allow repairs to the pipeline and asked for an increase in transit fees before agreeing to permit repairs. There are compelling reasons to speculate that the "accident" that damaged the pipeline may have been instigated by Libya, and perhaps even greater reasons to speculate that the stalling in repairs by the Syrians was in collusion with the Libyans. Suppliers to Europe hurriedly shifted to the Arab peninsula and Persian Gulf states, but this threw an enormous burden on already inadequate tanker fleets, which had to sail around the African continent to reach Europe. Consequently, freight charges between the Persian Gulf and Europe went from approximately $1.10 to $3.00 a barrel within four months (Mikdashi 1972, 148). The Libyan bargaining position benefitted immensely from this development. Agreement for a price increase was reached between the Libyan government and Occidental, which was permitted to step up production to its prenegotiation level. Other companies arrived at price increases also, first the smaller independents and finally the majors.

The Libyan episode demonstrated that the industrial nations are vulnerable in the short run to disruptions in oil supply, that price increases can be achieved by withholding supplies under market conditions favorable to the suppliers, and that the invincibility of the oil companies was a myth.[2]

The success of the Libyan settlement was followed by Iran's demands, which the U.S. government could not ignore. The Shah of Iran had developed a partnership with the United States to protect U.S. interests in the Persian Gulf. In November 1970, Iran negotiated a 5 percent increase in the tax rate and a nine-cent-a-barrel increase in the posted price of oil. This pattern was soon followed by other OPEC members; the year 1970, therefore, marked a turning point in the history of the oil industry and provided an impetus for OPEC to further better its terms of trade. In this effort they had the political and emotional support of most other developing nations, which depend to varying degrees on commodity exports for foreign exchange earnings to finance their development plans. They saw OPEC's vanguard actions as crucial to the commodity price debate on which the subsequent North-South dialogue came to be firmly hinged.

While these changes were taking place in the control of oil production and pricing, the world's dependence on oil as a source of energy was growing. There were warnings of impending disaster from time to time, but the oil companies and consuming nations chose to ignore them. For instance, James Akins (1973), one time U.S. ambassador to Saudi Arabia, warned consuming nations that a loss of production from any

---

2. For a fuller discussion of these past developments please see Fereidun Fesharaki and David T. Isaak, *OPEC, the Gulf, and the World Petroleum Market* (Boulder, Colo.: Westview Press, 1982).

two Middle East countries could cause panic among consumers and that the Arabs could use oil as a political weapon. But the market kept tightening all through the Spring of 1973, and prices were moving up. In June, OPEC called a meeting in Geneva, where it negotiated with oil companies for a 12 percent increase in oil prices (based on devaluation of the dollar).

In the ensuing months, political leaders of the Arab nations grew more vocal in their threats to use oil as a weapon against those countries that supported Israel. Again, Libya adopted the most threatening posture, and while celebrating the fourth anniversary of his revolution, Colonel Qaddafi announced the nationalization (51 percent control) of all oil companies operating in Libya. This was soon followed by the announcement of a price increase of Libyan oil to six dollars a barrel, almost twice the price of Persian Gulf oil. Meanwhile, OAPEC members grew more militant in their threats to use oil as a weapon, which led to a public warning by President Richard Nixon on the dangers and inevitable failure of an oil boycott, which Libya and other nations had been talking about.

In October, OPEC met once again with oil company delegates in Vienna. Just before the meeting, Egypt and Syria invaded Israel to recover the lands captured by Israel in the 1967 war. Arab representatives attending the meeting were incensed at the reports of U.S. arms and equipment being supplied to Israel, and in the prevailing mood of belligerence and shifting power in the market, agreement was elusive. OPEC demanded a posted price of six dollars a barrel, and the oil companies offered nothing more than a 15 percent increase. After a flurry of communications between the capitals of the Middle East oil countries and Washington, positions hardened swiftly. At Kuwait, OPEC announced a five-dollar-and-twelve-cent price per barrel; OAPEC announced an immediate 5 percent cut in production plus a 5 percent reduction each month, until Israel withdrew and specific rights of the Palestinian people were granted. OAPEC's position was buttressed by Saudi Arabia's subsequent announcement of a 10 percent cut in production and a total embargo in supply to the United States and the Netherlands. This was the last nail in the coffin of oil company power in the Middle East. Subsequently, OPEC emerged as the dominant power in the world oil market and assumed full control of pricing decisions.

In the immediate aftermath of the quadrupling of oil prices between 1973 and 1974, most analysts were incredulous—events appeared to defy rational explanation. OPEC itself, apparently, was quite surprised at the rapidity of developments and the results achieved. But soon thereafter came a spate of commentaries and analyses predicting a decline in prices after the Arab embargo was called off. Some researchers developed quantitative models of the oil market predicting a collapse of OPEC by 1980 (Kennedy 1974). Paul Lee Eckbo studied the record of commodity cartels in the past and arrived at the conclusion that "OPEC has also many of the characteristics of the earlier international commodity cartels that were successful for limited time periods" (1976, 108). This limited time period was four to six years for the earlier cartels. On this basis, OPEC was expected to collapse by the end of the seventies, but Eckbo's analysis was insightful enough to predict that "some successful post-OPEC cartel-like organizations" would take hold of the oil market beyond the OPEC era, and that the economic incentive to reestablish OPEC would be stronger than in the case of other known cartels. One cannot fault these well-reasoned studies for their predictions' not coming to pass, because OPEC's survival has been a surprise even to some of its own members. But there have undoubtedly been some monumental fallacies and simplistic assumptions at the root of most models depicting OPEC's structure and decision making. Typically, member nations of OPEC were equated with profit-maximizing firms operating as a cartel in a particular industry, a seriously faulty view. With the evidence now available, more plausible models can be put forward and examined.

## The Structure of OPEC

Through widespread usage, the term cartel has come to be viewed as almost synonymous with OPEC. In current literature on the world oil market, economists and others refer constantly to the OPEC cartel. Indeed the oil market of today—the recent drop in OPEC's share in total output notwithstanding—has many features that characterize the operation of a cartel. A cartel is formed when oligopolistic suppliers jointly agree to limit competition for the overall benefit of the group. The major incentive to cartelize comes from the prospects of increasing profits through the exercise of monopolistic power (or oligopolistic power, which is a diluted variant of monopoly), which suppliers can achieve only through collusive and concerted action. Most countries have some form of legislation making the exercise of oligopolistic power by a set of firms illegal, but in the international arena a set of sovereign states can vary well cartelize with impunity and without fear of legal violations.

The breakup of a cartel is predicated on the long-run effects of pricing decisions implemented by it. Figure 4.1 shows the market forces acting on a cartel. Under conditions of perfect competition, the quantity

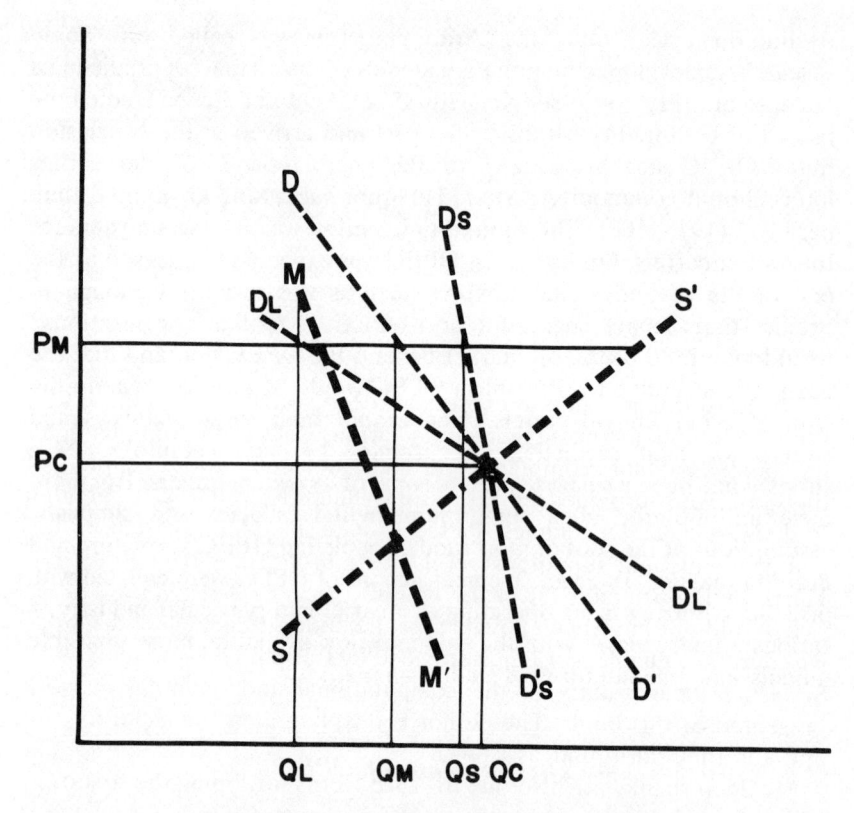

Figure 4.1 Market Forces Acting on a Cartel

produced and consumed is $Q_c$ and the equilibrium price, $P_c$. If the suppliers of the commodity were to form a successful cartel, they could theoretically limit output to the point where their marginal revenue curve $MM'$ intersects the marginal cost curve (the supply curve $SS'$) and raise prices to the point on the demand curve $DD'$ at which the market would be cleared. This price is shown by $P_M$ in figure 4.1. Using the industry supply curve $SS'$ and the demand curve $DD'$, the new price $P_M$ would lead to excess capacity $(Q_C - Q_M)$. In actual fact, however, in the short run, demand is highly inelastic for a commodity like oil, and the change in quantity demanded may not travel along the demand curve $DD'$ but the short-run demand curve $D_S D'_S$, resulting in a very small excess capacity of $(Q_C - Q_S)$. Thus producers would have to cut back on supplies by a very small quantity in the immediate short run, and profits could increase substantially.

Given time for changes on the demand side, however, the quantity demanded would reduce to $Q_M$, *ceteris paribus*, and the excess capacity would increase to $(Q_C - Q_M)$. In fact, if the economy is permitted an even longer time, other substitutes may be developed, technological innovations may occur, and oil-using capital stock may be changed to reflect these changes, making the longer run demand curve even more elastic, as shown by $D_L\ D'_L$. If this happens, then at the price $P_M$, quantity demanded would drop even further, to $Q_L$.

The forces that typically bring about the breakup of a cartel arise from the excess capacity depicted by $(Q_C - Q_M)$ and $(Q_C - Q_L)$. Under these conditions, the incentive for a member to increase output at the prevailing price is very strong. The quota system then breaks down, and an excess supply at the cartel price ensues, exerting downward pressure on the price, leading ultimately to the breakup of the cartel. The strength of a cartel, therefore, rests critically on the extent to which its members adhere to the quota system.

In the case of OPEC, it is not enough to say that since it has survived as an effective market force it is not a typical cartel. The reasons for not treating OPEC as a cartel go beyond evidence of its success and are derived from an analysis of its structure. Among many studies dealing with this subject, the work of Edward W. Erickson (in Pachauri 1980 b) appears the most persuasive and convincing. His analysis explains OPEC's structure and behavior by treating it as a dominant-firm price-leadership model, with Saudi Arabia as the dominant firm. The recent history of oil price and quantity changes shows clearly that Saudi Arabia, which is the largest producer within OPEC (and therefore the low-cost producer), has dominated OPEC's decision making.

Typically, as Erickson explains, dominant-firm price leadership also implies a restriction in output, but as opposed to a cartel, the restriction is applied passively, resting on a very loose coordination of output decisions among members. Slack in the market is taken up by the dominant firm by effecting changes in ouput. This feature has been in evidence with credible consistency in the post-1973 oil market. Table 4.1 shows changes in the output of OPEC members since 1973. Saudi Arabia's output, except for 1979–80, changed in the same direction as that of total OPEC output, while this was not the case for all members. Further, changes in Saudi Arabian output generally constituted over 40 percent of total OPEC changes, as can be seen in the last row of table 4.1.

There have been periods in the recent past when Saudi Arabia's role may have been subordinated to the pressure of the more "hawkish" members of the group, who seem to regard the Saudis as too soft in the matter of raising prices. But given the volatility and rhetoric of Middle East politics, it is difficult to arrive at definite conclusions from public postures. Consequently, whether the oil price increases of 1979–80,

## Table 4.1

Changes in Annual Oil Output Since 1973, OPEC Countries, 1975–81
(thousands of barrels)

| Country | 1975–76 | 1976–77 | 1977–78 | 1978–79 | 1979–80 | 1980–81 |
|---|---|---|---|---|---|---|
| Saudi Arabia | 556.8 | 218.6 | −328.0 | 447.2 | 156.6 | − 45.8 |
| Iran | 201.0 | − 86.4 | −169.6 | −794.6 | − 551.2 | − 91.9 |
| Kuwait | 26.0 | − 66.7 | 45.3 | 151.4 | − 311.6 | − 194.2 |
| Iraq | 53.8 | 35.6 | 49.6 | 292.8 | − 263.9 | − 653.6 |
| United Arab Emirates | 92.1 | 24.3 | − 66.6 | − 0.2 | − 44.7 | − 76.3 |
| Qatar | 17.8 | − 19.0 | 16.8 | 8.9 | − 12.3 | − 24.8 |
| Libya | 148.3 | 58.5 | − 36.5 | 32.0 | − 99.0 | − 237.3 |
| Algeria | 40.2 | 26.3 | 28.8 | − 1.2 | − 41.4 | − 29.4 |
| Nigeria | 106.1 | 7.4 | − 69.0 | 144.4 | − 86.9 | − 227.8 |
| Gabon | 0.0 | − 0.4 | − 4.8 | − 2.2 | − 10.5 | − 9.0 |
| Ecuador | 9.6 | − 2.1 | 7.5 | 4.4 | − 3.5 | 2.2 |
| Venezuela | − 22.1 | − 16.3 | − 26.6 | 139.7 | − 69.4 | − 23.6 |
| Indonesia | 68.3 | 65.8 | − 17.8 | − 16.1 | − 4.4 | 8.2 |
| Total | 1,297.9 | 245.6 | −570.9 | 406.5 | −1,342.2 | −1,603.3 |
| Ratio of Saudi Arabia change to total (%) | 42.9 | 89.0 | 57.4 | 110.0 | − 11.7 | 2.8 |

Source: *Petroleum Economist*, various issues.

which followed events connected with the Iranian revolution, were forced on a reluctant Saudi Arabia by the so-called price hawks of OPEC is a subject for speculation only. Certainly, analysts who link oil developments in the Middle East to utterances of national leaders are treading on slippery ground: these utterances provide little insight into actual decision making. A more meaningful analysis would be based on developments in OPEC countries related to prices and quantities in the oil market, such as output changes, price discounts (offered by most members from time to time), and internal economic development. Two issues require discussion in this context. First, the behavior of OPEC cannot be viewed merely as the actions of a group of profit-maximizing firms in an industry. Even though profit maximization is a dominant OPEC motive, it is often relegated to a lower priority when political considerations demand. Second, the oil market has strong influences not necessarily controlled by OPEC. Developments on the demand side have proved to be of greater importance in determining market changes than anticipated. Then there is, of course, sizable production outside OPEC, which exerts considerable pressure on the market.

The clearest manifestation of the weakness of OPEC's profit-maximization motive is OPEC's almost fanatical moves to prevent downward trends in oil prices even when they would maximize profits over the medium or long term. It is difficult to believe that the evidence of long-run price elasticity of demand and price elastic supply by non-OPEC producers could possibly have been lost on OPEC. Indeed, Saudi Arabian authorities have made no secret of their concern about the likely substitution of other oil for OPEC oil resulting from rapid price increases.[3] Yet, the pre-OPEC decline in prices, particularly the large reduction of 1959, appears to have left an indelible mark on OPEC, which props up sagging prices without apparent rationale, as events in early 1982 show. According to Zuhayr Mikdashi (1972, 97), all this is, of course, in keeping with the objective stated in the very first resolution at the OPEC conference: "Members shall study and formulate a system to ensure the stabilization of prices by, among other means, the regulation of production." Mikdashi also states that the aim of the originators of OPEC's statutes was not to constrain OPEC to a single objective, which would become obsolete over time, but to ensure a continuous process of consultation, cooperation, and exchange of information. In other words, they felt that OPEC could be kept alive only if several goals could be sought in its self-interest and new goals added or old ones replaced from time to time.

A major reason for OPEC's success has been its flexible approach, the lack of rigid quotas, and a decision-making process that allows its members to express their views. These are possible because of Saudi Arabia's enormous influence on the market. Saudi Arabia would not jeopardize that influence by following a path detrimental to the overall interests of OPEC unless there were a serious conflict of interest between Saudi Arabia and the others. M. A. Adelman (1981, 7) underlines the pivotal Saudi role in pricing and output restrictions immediately after the Iranian revolution. In Adelman's view, the oil shortage in early 1979 was not caused by the Iranian revolution but by lack of capacity elsewhere. The tightness of the market did not reach panic proportions until the deliberate Saudi cut from ten to eight million barrels a day, which created a world crisis in February 1979. The resultant higher price stabilized in the spring; then Saudi Arabia cut

---

3. In reply to a question at a recent seminar, Sheikh Yamani is reported to have said, "The sharp increases in the price of oil which took place during 1979 and 1980 did affect demand—they have a short-term effect and a long-term effect. Probably the long-term effect has been greater than the short term. . . . Coal is the main competitor for oil. It is taking a good portion of the oil market, or the fuel oil segment. . . . My own view is that if OPEC maintains the present price of oil in money terms for at least until the end of 1982, this in itself will stimulate demand. And if the increases in the price of OPEC oil in the period between 1982 and 1986 will be very reasonable, very nominal, then definitely demand for oil will increase no matter what are the huge investments in the field of conservation and substitutes" (*Middle East Economic Survey*, supplement to vol. 24(51) 6–7).

output again, which brought about another upward spurt in prices. According to Adelman, Sheikh Yamani termed the 1979–80 price explosion as "another corrective action." Apparently, Saudi domination of and complicity with OPEC works in the opposite direction as well. The relatively high Saudi output levels of late 1981, almost in defiance of the price hawks in Libya and Algeria, were also in OPEC's interest. Higher prices at a time of recession in the industrial world would have jeopardized the value of Saudi and Kuwaiti assets in these countries and provided impetus to energy conservation measures.

The sensitivity of OPEC to market changes in a period of slackening demand can be seen in figure 4.2. Line $D_1 D'_1$ represents the global demand curve for oil at any given point of time; $S_o S'_o$ is OPEC's supply curve; and $S_w S'_w$ is the global supply curve, which is essentially the summation of supply curves of OPEC and non-OPEC producers. Line $S_o S'_o$ is less price elastic than $S_w W'_w$ because it is assumed that non-OPEC suppliers would be more responsive to price changes than OPEC, which sets prices collectively through the exercise of oligopolistic power. In fact other analysts generally depict the supply curve for OPEC as backward bending (Ezzati 1978, 133–36).

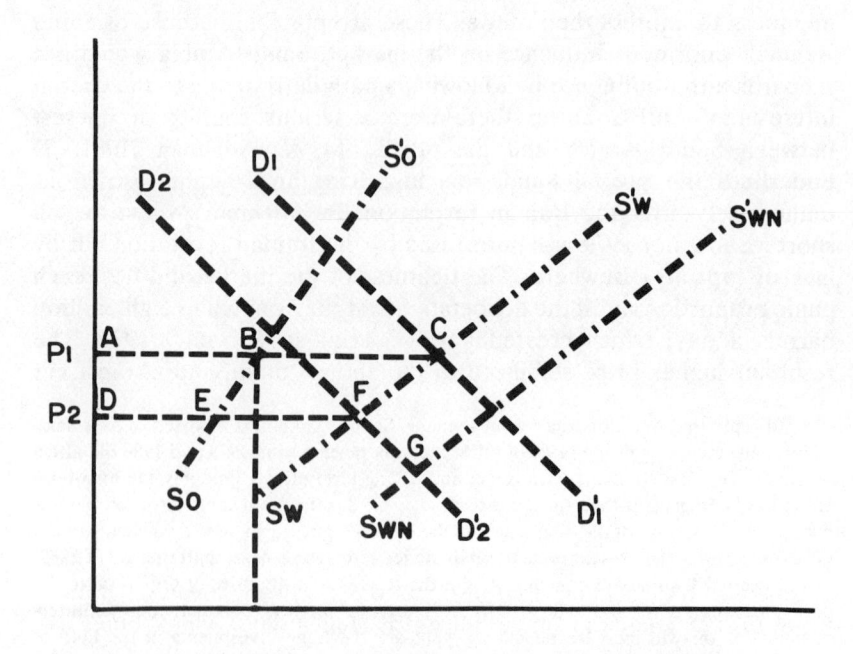

Figure 4.2 Sensitivity of OPEC to Market Changes

The market is initially in equilibrium at a price of $P_1$, with quantity $AB$ being supplied by OPEC and quantity $BC$ being supplied by non-OPEC producers. If over a period of time the demand curve shifts to the position $D_2D_2'$ as a result of downward income changes (recession) and of technological changes in the supply and use of substitute sources of energy, then the new equilibrium price changes to $P_2$. At this price, OPEC would be supplying a quantity $DE$ and non-OPEC producers, $EF$. The new price $P_2$ would, of course, be reached with an initial reduction by non-OPEC producers responding to the changes in demand. If OPEC does not drop prices at the same time, consumers would shift from OPEC to non-OPEC sources at prices even marginally lower than $P_1$. Correspondingly, the reduction in demand for OPEC oil would result in pressure for lower prices. Ultimately, OPEC would lower prices to $P_2$, at which the market would clear.

While a shift in demand takes place, if an accompanying outward shift in the supply curve also occurs to, say, $S_{WN}S'_{WN}$, then the pressure for downward movement in prices would increase further, tending to lead to the new equilibrium quantity and price represented by $G$. Price setting by OPEC can, therefore, hold only if they correctly anticipate downward shifts in demand as represented by $D_2D_2'$ or outward shifts in supply as represented by $S_{WN}S'_{WN}$. The developments of 1981–82 throw some doubt on whether OPEC's predictions took these into precise account in raising oil prices during 1979–80. Quite naturally, a reduction in prices and quantities for OPEC exports would introduce an element of instability into the group's collective actions—the larger the reduction in exports and real prices, the greater the instability. During these periods, Saudi Arabia would assume predominance on account of its ability to soften the blow for other members, whose revenues could otherwise seriously erode. As a result of these changes, OPEC's price-setting leadership in the oil market would be likely to undergo changes from time to time, depending on OPEC's share of total world consumption, and Saudi Arabia's ability to cut production, and its share in total OPEC production.

A simple OPEC stability index ($SI$) can be computed as

$$SI = \frac{S_o}{S_w} \left[ \frac{S_A - 6}{12} \left( \frac{S_A}{S_m} \right) \right],$$

where $S_o$ is total OPEC output of oil, $S_w$ is total world output of oil, $S_A$ is total Saudi Arabian output of oil, and $S_m$ is mean country output of other OPEC members (in millions of barrels a day). The value of the index would be unity when OPEC output was approximately 50 percent of world output, and Saudi output was at a peak capacity of 12 million barrels a day, and the mean output of other OPEC producers was 25 percent of Saudi output. At Saudi production levels of less than six million barrels a day, the value of the index would be negative, indicating a high degree of instability.

Values of the stability index have been computed and plotted in figure 4.3 for the period 1973 to 1982. It can be seen that the index rises and

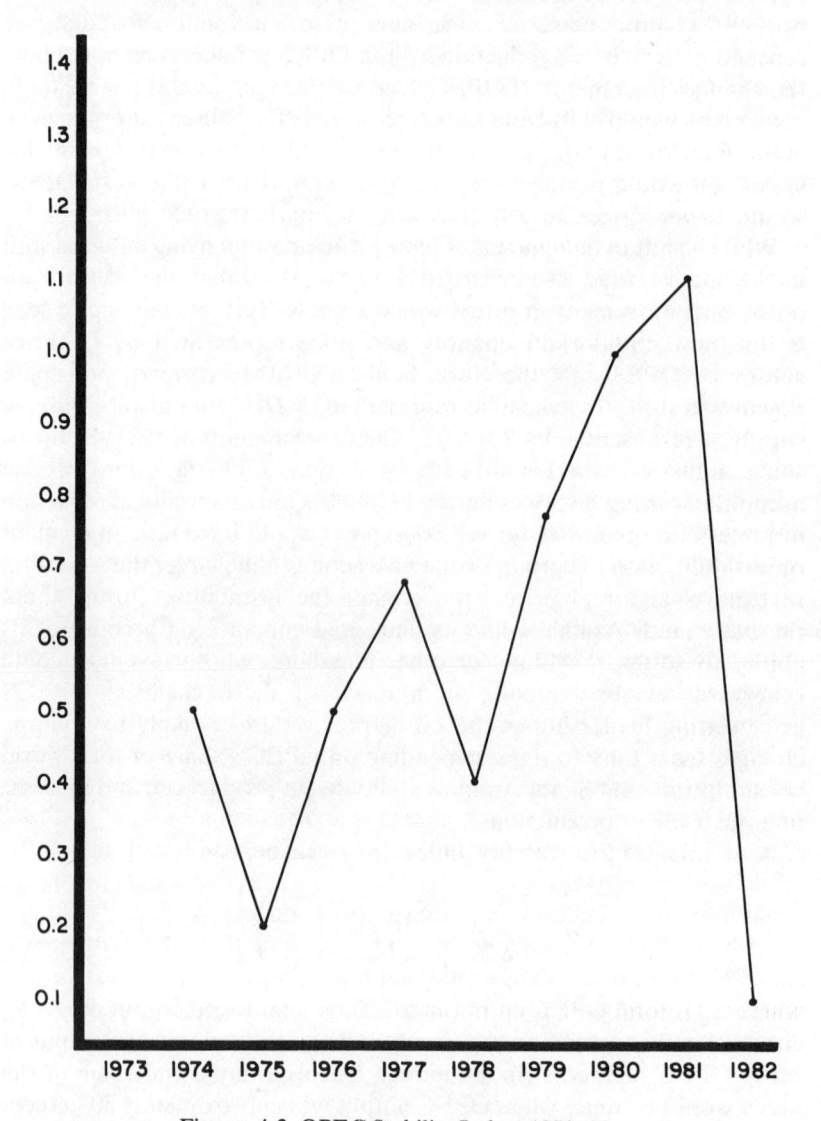

Figure 4.3 OPEC Stability Index, 1973–82

falls with apparent changes in OPEC's power, and that 1982 was the worst year for OPEC's stability, as events appear to corroborate. But these changes are relative and provide no measure of absolute instability, nor do they lead to predictions of OPEC's breakup. There is every reason to believe that OPEC is robust enough to withstand both external and internal strains and will remain a force to reckon with in the world's energy future.

## OPEC Objectives and Internal Dynamics

OPEC's flexibility is not evidence of an absence of internal conflict, and in fact there is a constant current of conflict among its members, an extreme example of which is the war between Iran and Iraq and a less violent one, the rhetoric of Libya's Qaddafi against Saudi Arabia's rulers. But these skirmishes are not new, and OPEC's business has gone on in spite of them. There are, of course, major economic and political disparities among members of OPEC (discussed in detail subsequently), especially the Middle East. In a region where power balances can shift very quickly, alliances and animosities are changing constantly: Egypt and Libya's brief honeymoon during the short existence of the United Arab Republic changed to a virulent hostility on the part of Libya's Qaddafi toward Egypt's Anwar Sadat; the 1979 union of Syria and Iraq changed by 1981 to Syria's open support for Iran in the Iran-Iraq war; and of course the Iran-Iraq war itself is extreme evidence of changing relations within OPEC.

Apart from shifts in alliances and hostilities, which are often unpredictable, certain Middle East members of OPEC also have philosophical and political differences of a more lasting nature. For instance, Qaddafi was originally opposed to the inclusion of monarchies such as Saudi Arabia in any future federation of Arab states, because "monarchies had delayed the process of Arab unity" (Mikdashi 1972, 91). Qaddafi's early 1982 condemnation of the Saudis as the enemies of Arab nationalism and the surrogates of western imperialism in the Middle East was merely a reiteration of this deep-seated suspicion. The monarchies of the region, on the other hand, are apprehensive about the dangers that revolutionary governments pose to their own conservative societies and systems of government. The Libyan regime is, therefore, looked on with considerable annoyance and contempt by the princes of Saudi Arabia; and until a closer understanding had developed by the beginning of the eighties, revolutionary Iraq and monarchical Saudi Arabia had a troubled relationship.

Even influences that unite the region cause sharp divisions. There is, first, the common enemy Israel, which the Arab nations solidly (and even Iran in rhetoric) are vowed to fight almost on a permanent basis. But common anti-Israel policies do not translate into common security measures or unity. The so-called front-line states of Egypt, Syria, Jordan, and Lebanon have widely differing perceptions of their burdens in the anti-Israel effort, justified by the damage borne in past wars. This leads to feelings of discrimination and differential risk among the front-line states, which weary of carrying the heavy burdens in furtherance of a common cause inequitably championed. The Camp David Accord continues, even after Sadat's death, to receive the support of the Egyptian people, who prefer peace to the dubious distinction of being the greatest sufferers in the struggle against Israel. U.S. efforts toward a strategic consensus among moderate Arab states, even if partly successful, could further separate the moderates and the radicals. The radicals, for instance, are impatient at Israel's resistance to forming a Palestinian state on present Israeli territory, and they see moderates like Saudi Arabia hindering their avowed all-Arab goal.

The second contradiction in Middle East relationships is the way OAPEC's evolution and internal developments affect OPEC's history and internal developments. The founders of OAPEC, Saudi Arabia, Kuwait, and Libya, initially intended membership to be limited to those Arab states in which oil was "the principal and basic source of its national income" (Mikdashi 1972, 104). Since this stipulation does not appear definitive, various other countries have joined OAPEC: Iraq in 1968, and Algeria, Abu Dhabi, Bahrain, Dubai, and Qatar in 1970. Arab countries are not automatically eligible for membership, which requires the unanimous acceptance of the three founding members. Unlike the Arab League, OAPEC excludes non oil-exporting countries of the Arab world, so it is very much a group with vital interests in the world oil market. OAPEC is, therefore, a subset of OPEC as well as of the Arab League, and by virtue of its large oil revenues, exerts great influence in both these forums, and its internal strains carry over into the deliberations of both these bodies.

However, OAPEC's charter decrees that its policies and activities must not be at variance with those of OPEC and further stipulates that OPEC decisions are binding on all members of OAPEC, even those who are not members of OPEC. Each OPEC decision has been faithfully upheld by members of OAPEC, a record that has helped allay the doubts of non-Arab members of OPEC, such as Iran, Indonesia, and Venezuela.

At the time of its birth, OAPEC was a group of monarchies, since at that stage Libya, too, was a kingdom. But its makeup has changed with the revolution in Libya and the addition of Iraq and Algeria, which have socialistic forms of revolutionary governments.

The pressure groups that influence OPEC decisions both internally and externally have to work within clearly specified procedures and mechanisms. The highest forum for decision making is the ministerial conference, where all major pricing decisions have been either ratified or made. At each meeting of the conference, member countries are represented by delegations, with one designated leader responsible for the positions and votes of that particular country. The conference meets twice a year in the ordinary course, but special meetings have been called from time to time by a simple majority of member nations endorsing any individual request to the secretary-general. The duties and responsibilities of the conference as laid down in Article 15 of the OPEC charter, include (1) formulating the general policy of the organization, (2) considering new applications for membership, and (3) approving the appointment of members of the board of governors.

It is significant that although the resolutions of the conference founding OPEC in 1960 are centered entirely on increasing oil prices and ensuring stable and growing revenues for its members,[4] the articles describing its mode of operation are general and could very well apply to the functioning of a social club. This implies that the founding members of OPEC were confident that the credentials of countries admitted to OPEC, the importance of oil revenues to their economic progress, and the preeminent power of some founding members in the global oil market were enough to ensure the focus of the group. There is reason to believe that the larger producers, notably Saudi Arabia and Iran, preferred an unstructured approach to decision making when establishing the organization, thus ensuring their own power in its functioning. The founding resolutions are, therefore, to be seen only as the organization's underlying philosophy and not necessarily as a blueprint for action. For instance, Resolution 1.1 mentions the use of "regulation of production" as a possible means for stabilization of prices, but no explicit quotas have ever been strictly enforced. This again has been possible due to the dominant market position of Saudi Arabia and its ability to accept cuts in production out of proportion to OPEC's overall production cuts.

Political divisions and tensions thus far appear to have had only a minimal impact on consensus decisions within OPEC. Nor has there been any serious dispute between this body and OAPEC. Members of OPEC seem to accept political strains, perceiving them as lying outside

---

4. Mana Saeed Al-Otaiba, *OPEC and the Petroleum Industry* (New York: Wiley, 1975), pp. 58–60.

their collective self-interest. On the other hand, economic disparities and strains are probably of greater consequence, and where these economic divisions are reinforced by political divisions, the impact could be unfavorable for OPEC's unity.

Ali Ezzati (1978) has examined the internal factors that affect OPEC strategies and lists the major determinants of bargaining power of each OPEC country. These include absorptive capacity (for effectively utilizing oil revenues), population, economic infrastructure, production costs, commitments to oil companies, and political and social factors. Ezzati emphasizes that even though OPEC effectively maintained unified pricing and production policies in the past, economic and political differences among members may not permit the same stability to continue in the future: one country's optimum price and output strategy may not hold for another.

Ezzati's analysis is representative of new research, which bypasses the traditional neoclassical view of OPEC as a cartel with each member maximizing economic rents (the difference between market price and marginal cost of production). The "bean counting" approach advocated by Fereidun Fesharaki (1981) also emphasizes the limitations of viewing members of OPEC as profit-maximizing firms exercising oligopolistic power. He is rightly critical of the popular but mistaken notion that OPEC will meet the residual demand for its oil after the production from non-OPEC sources has been accounted for in the total world demand for oil. According to Fesharaki, a realistic approach consists of examining the technical, economic, and political factors governing production decisions. The assumptions and approach of these new researchers are at variance with the views of most oil companies in the seventies and the work of distinguished researchers who use traditional quantitative models in depicting OPEC behaviour. Most of these models are based on individual country supply curves, which are upward sloping and have significantly high price elasticities. These models are partly valid for high absorbers within OPEC but certainly do not apply to the dominant low absorbers.

The absorptive capacity of each OPEC member is the most important economic factor governing its production strategy. This capacity is normally defined as gross domestic investment as a percentage of gross national product in any period that would, under existing conditions, provide an acceptable rate of return. But the concept goes further than a cut-off rate of return on investment and includes undesirable social and political effects of investments. In order to avoid these effects, a country may deliberately limit production to levels that are in consonance with balanced growth and planned development. A case in point is the oil export policy of Mexico, which calls for export levels far below its full capacity and at prices which undoubtedly are, and will remain, well above marginal costs (Landsberg 1979, 252). Thus a detailed and comprehensive analysis of the sociopolitical and economic conditions in each country is necessary for a reliable assessment of oil export strategies. In the following chapter such an assessment has been undertaken for selected Middle East countries. But a few issues pertaining to OPEC's stability remain to be explored here.

There are serious implications for OPEC's stability arising from the attitudes and actions of the major consuming nations, who not only look with disfavor on the loss of oil company control in the major producing regions of the world but also resent the economic jolt OPEC's price increases imposed on the economies of the western nations since 1973. Even though the oil companies no longer dominate oil production in the OPEC countries, they still have the major share of refining, marketing, and distribution activities in the noncommunist nations. Their influence in certain regions is by no means negligible, and even in countries where they no longer own oil-producing assets, they continue to provide technical expertise and other services. In periods of growing intra-OPEC strains, therefore, oil companies can force a country to sell at less than the established OPEC price, either directly or by not buying in adequate quantities from that country. Quite naturally, a non-Arab OPEC member that is a high absorber is the most vulnerable target. The economic pressure on Nigeria was predictably disturbing to other members of OPEC, particularly in the period of slack oil demand in early 1982, and suspicion that oil companies were advising Nigeria to break away from OPEC ran high among other oil producers, particularly in the Middle East.[5]

In 1984 the turmoil of the world oil market is still not behind us. The post-1981 oil glut and underutilized production capacity is only a temporary reprieve—a short period of adjustment in the long-run scarcity of hydrocarbon fuels. The past two years have tested OPEC, and doubts about its ability to continue in existence are quite justified. The imposition of production quotas for individual members in March 1983, which led to a price cut for the first time in OPEC's history, gladdened the hearts of analysts who believe that production quotas in a period of slackening demand sound the death knell of OPEC. Even though the quotas, after intensive and hard bargaining, were accepted, there have been reports that some members continued producing above their ceilings. How long the uneasy agreement holds depends, in the short run, on how fast global demand picks up.

5. "Demand Slump Restrains Global Oil, Condensate Output in 1981," *Oil and Gas Journal*, March 8, 1982, pp. 99–103.

But these recent developments do not in any way refute the thesis that Saudi Arabia is the dominant producer within OPEC and has an overwhelming influence on its actions. Its role as "swing producer" underlies the 1983 production quotas and their fragile impact on the unity of OPEC. For instance, Saudi Arabia's output in the first quarter of 1983 averaged 3.9 million barrels a day against a ceiling of 5 million barrels a day, whereas in the same period Iran produced 2.45 million barrels a day against a ceiling of 2.4, and Venezuela 1.9 million barrels a day against 1.675.[6] It is easy to visualize OPEC's major nightmare—Saudi Arabia increasing its output in the face of a worldwide glut and forcing a decline in prices. Such a prospect would inevitably lead to the collapse of production agreements and to OPEC's demise. No OPEC member is naive enough to deny the enormous market power that this possibility bestows on Saudi Arabia, a power that would not decline materially when demand increases and there are further negotiations on revised quotas. Such an event would, of course, result in further bickering between the high and low absorbers, requiring Saudi Arabia to make concessions (perhaps from a position adopted in anticipation of such concessions) to bring about agreement.

The longer low oil revenues continue, obviously, the deeper will be the disparities between the high and low absorbers. There is already a backlog of planned development projects not undertaken due to reductions in oil revenues. The members of OAPEC are likely to suffer a reduction in oil revenue of around $117.68 billion for the years 1982 and 1983.[7] These reductions would, of course, have been larger had they not been cushioned by an accretion in the strength of the U.S. dollar. This fact has sheltered OPEC from the full decline in prices and boosted its purchasing power, particularly in Japan and Western Europe. According to one estimate, in April 1983, soon after the oil price cut, a barrel of oil bought OPEC only 8 percent less in manufactured goods from industrial countries than it had in late 1980, despite a large drop in oil prices and general inflation in the west.[8]

The past two years have provided an opportunity for a re-appraisal of development strategies among the members of OPEC. Recent writings and pronouncements indicate that a shift toward a new strategy of oil-exports economic-development is likely in the future. This policy is expected to consist of (1) an emphasis on development of the nonoil

6. *Petroleum Intelligence Weekly*, March 21, 1983, p. 5.
7. *OAPEC Bulletin* 9(5), 1.
8. *Petroleum Intelligence Weekly*, April 11, 1983, p. 2. Given this fact, Iran's reported discount in oil supplies to Japan, even though a violation of the OPEC agreement, does not appear all that irrational and is an attempt to change the price of oil from dollar values to valuation based on purchasing power in the international market.

sectors of the economy; (2) a curb on imports of consumer goods and greater import substitution, aiming at a higher degree of self-reliance; (3) a shift to fewer overseas investments and in some cases a disinvestment; (4) a lengthening of the life of hydrocarbon resources through lower production rates; and (5) a less ambitious economic development. In general, there is likely to be a close correspondence between oil revenues and more austere development needs.

But such a common policy of less extravagance in expenditure, unfortunately, would not reduce the strains that have undermined OPEC's strength in recent years. The unexpected weakening of oil markets since 1979–80 does raise doubts on the future strength of OPEC. And OPEC's internecine squabbles have even stronger implications for its future. For instance, an Iranian takeover of Iraq's oil installations is fraught with danger to OPEC and indeed to the entire oil market, although Adeed Dawisha discounts such a possibility on the grounds that Iran's goals are religious and not economic and that a major Iranian objective of the war is the overthrow of Saddam Hussain (Stock 1982, 281–82). Furthermore, the threat of domination by Iran's Shia theocracy is a source of considerable fear to the Arab states of the region, and the threat of violent action by Shia fundamentalists is felt even in Saudi Arabia's eastern region (which is populated by a large number of Shias), particularly after the attempted coup in Bahrain in 1982. The Gulf Cooperation Council was formed because of the anxiety of Arab countries in the Persian Gulf and their need to present a cohesive front to Iran. Ostensibly, the council's objective is to bring about economic and financial integration of the member states, but their collective security and defense are certainly a focus of their activities and deliberations.

CHAPTER 5
# OIL REVENUES: DETERMINANTS AND IMPACT

Oil output and export decisions made by OPEC are best assessed within the framework of individual country analysis. Such analysis must include an assessment of the social, economic, and political dimensions of national development and the role played by oil revenues in these dimensions. It need hardly be mentioned that a study of the oil sector in each country is necessary, but this in itself is not sufficient. In effect, the division of OPEC nations into high and low absorbers should be carried further—the benefits and costs of every dollar of oil revenues require assessment in the broader context of economic, political, and social effects.

In discussing these questions in this chapter, I concentrate on a small number of Middle Eastern countries, since the stability of OPEC as a whole is largely dependent on the objectives and motivations of these countries. We have seen earlier the influence of Saudi Arabia. This kingdom is a member of the community of Middle Eastern Islamic countries, and thus its postures and actions cannot be isolated from the common aspirations, strengths, weaknesses, and contradictions of that group. It is naive to believe that any of these nations would act permanently against the community's self-interest, or that any outside nation through diplomacy, sales of military hardware, or appeals for restraint in oil price increases, would get any of these nations to do its bidding. M. A. Adelman's (1981, 12) sarcastic reference to the U.S. "special relationship" with Saudi Arabia being as useful as its "special relationship" with Iran underlines a profound reality, namely, that the expediency of such policies is perhaps detrimental to the long-run interests of all oil-consuming countries and certainly an example of naivety in international economic relations. At the same time, it would be myopic to assume that the nations of the Middle East are unable to evaluate and apply their own self-interest in the oil business. OPEC undoubtedly has made serious errors, particularly its price increases of 1979–80, but it has been no worse than those of any other cartel. There is in fact evidence of remarkable expertise in policy analysis on various aspects of the world oil market.

The members of OPEC also realize the importance of oil in their own national development and destiny. A Kuwaiti who remembers $21 as the per capita income in current dollars at the end of World War II (Amin 1974, 5) cannot fail to notice the miracle of oil power, which has enabled that nation to reach a per capita gross domestic product of over $17,000 in 1979 (World Bank 1981*b*, table 1). Quite apart from changes in income, the growth of oil production and the infusion of oil revenues have brought about profound changes in the sociopolitical milieu and the structural characteristics of their economic systems. Even though some oil-exporting nations of the Middle East launched ambitious development plans before 1973, the pace of new development reached frenzied levels soon after the 1973–74 price increases. Many of these efforts seemed to be aimed at activating the nonoil sectors of their economies, so that these sectors would complement and subsequently replace income from oil exports. But a large portion of oil revenues were used to import goods and services for current consumption.[1]

Changes in OPEC's oil revenues are evident in figures presented in table 5.1. Revenues increased sharply following the 1973–74 price increases, then declined in 1975 and 1978. The next round of price increases in 1979 and 1980 brought another boost in earnings, resulting in an estimated current-account surplus of $108 billion in 1980 from $65 billion in 1979.[2] These figures do not reflect changes on the export side only; in 1979 there was a decline of 12 percent in imports (largely due to political changes in Iran), followed by a 16 percent increase in

TABLE 5.1

OPEC Oil Revenues (billions of dollars)

| 1973 | 1974 | 1975 | 1976 | 1977 | 1978 | 1979 | 1980 | 1981 | 1982 | 1983[a] |
|------|------|------|------|------|------|------|------|------|------|---------|
| 34.8 | 112.0 | 103.6 | 126.0 | 137.9 | 131.9 | 199.1 | 278.4 | 287.9 | 191.4 | 168.0 |

[a] Estimate.

1. While on a consulting assignment in Iraq some years ago, I learned that two senior Iraqi professionals in a training institution had gone to Italy on a "mission"—to buy office furniture for themselves and their colleagues (in a socialist state known for its philosophy of frugality!). This trivial anecdote is recalled only to emphasize the inefficiencies in expenditure when amounts of money are available totally beyond anticipated projections and for which procedures, institutions, and infrastructure are generally quite inadequate.

2. Bank for International Settlements, *Fifty-first Annual Report* (Basle, Switzerland: 1981), p. 96.

TABLE 5.2

Exports of Goods and Nonfactor Services, OPEC Countries
(Percent of gross domestic product)

| Country | 1960 | 1979 |
|---|---|---|
| Algeria | 28 | 32 |
| Ecuador | 17 | 24 |
| Saudi Arabia | * | 60 |
| Indonesia | 13 | 30 |
| Iran | 19 | * |
| Iraq | 42 | 63 |
| Kuwait | * | 69 |
| Libya | * | 70 |
| Nigeria | 15 | 25 |
| Venezuela | 32 | 31 |

Source: World Bank 1981b, table 5.
* Not known.

1980. Exports declined 13 percent in 1980, compared to a small increase in 1979.

A direct consequence of increased oil prices has been an upsurge in the ratio of foreign trade to national output. Table 5.2 shows the distribution of exports of goods and nonfactor services as a share of gross domestic product in OPEC nations for the years 1960 and 1979. The differences across countries are substantial, as are the changes within countries between 1960 and 1979. But the actual determinants of development in its most comprehensive sense vary substantially across nations and can be analyzed only in country-specific terms. In the following sections, therefore, we evaluate historic development in Algeria, Iran, Iraq, Kuwait, Libya, Saudi Arabia, the United Arab Emirates, Bahrain, Qatar, and Oman. The last four are considerably homogeneous in their economic, social, and political characteristics, even though Bahrain and Oman are not members of OPEC.

## Algeria

Algeria's approach to development and its political philosophy have much to do with its colonial past. Characteristic of most countries colonized by the French, Algeria exhibits strong French cultural influences upon a typical Arab society. Algeria occupied a special place in the colonial French empire, since France harbored ambitions of making Algeria an extension of France (a fantasy inspired by the geographical contiguity of the two nations). Consequently, independent

Algeria inherited a fairly adequate infrastructure but a society excessively dependent on the French for technical and administrative skills. French control extended even to ownership of agricultural land, with 28 percent of the most fertile land in the hands of Europeans.

The first year following independence in 1962 was one of uneven transition and serious economic dislocation. There was a sudden withdrawal of skilled personnel—all European nationals—and the entire Algerian economy suffered idleness and even sabotage. Productive assets changed ownership and brought about a redistribution of wealth. The emphasis on socialism retarded the development of entrepreneurial skills, which could have accelerated economic growth, but was perhaps instrumental in holding in check income disparities that often tend to grow in developing economies. Another significant development that followed independence was the creation in 1963 of Sonatrach, the organization responsible for the management of Algeria's oil industry. The government also set up a central bank to develop and implement the country's monetary policy.

Of particular importance to Algeria's economy since independence has been the slow growth in agricultural earnings. Agrarian reform in 1966 shifted land ownership to those who lived and worked on it. Land formerly held by the French settlers is now managed by committees and run as collectives. Disparities between the collectivized farms, which are relatively modern, and the privately cultivated traditional farms are very marked. Even though agriculture was the dominant sector in the preoil era, its growth since has been sluggish, with an annual rate of 0.4 percent between 1960 and 1970 and 0.6 percent between 1970 and 1979. Consequently, in 1960 agriculture contributed 21 percent of gross domestic product, but by 1979 its share had declined to 7 percent (World Bank 1981*b*, tables 2, 3). Land reform has, as evidenced by these figures, altered the institutional and structural aspects of agriculture but have had little impact on production. Agriculture in general is, therefore, in a position to absorb large inputs of resources, particularly considering that even by 1980, 56 percent of the population lived in rural areas.

Konrad Schliephake (1977) has investigated the impact of the oil industry on the Algerian economy and has found that not only have oil and gas development provided high employment, but the opening up of highways and transport has also had a major impact on regional development. The effect of oil revenues on agriculture thus far appears weak; its effect is more pronounced on trade, commerce, and industry. The value added in manufacturing (1975 dollars) increased from $967 million in 1970 to $2,220 million in 1978 (World Bank 1981*b*, table 3). Socialist philosophy formalized in the government's document "Charte de l'Organisation Socialiste des Enterprises" governs industrial development, which relies heavily on foreign loans and grants and, more recently, oil revenues. Most external aid has come from nations in the Organisation for Economic Cooperation and Development, the centrally planned economies, and the Kuwait Fund for Arab Economic Development. In 1977–78, Algeria had to borrow heavily from the Eurodollar market to finance development plans, but this borrowing declined from $3,297 million in 1978 to $2,039 million in 1979 (UNIDO, December 1979). New industrial projects launched in the 1970s are in the fields of gas liquefaction, petrochemicals, tractors and farm machinery, and heavy engineering and fabrication. Most industries are characterized by low operating efficiencies and are hampered by the lack of managerial and other industrial skills.

In general the Algerian economy has benefitted greatly from the infusion of larger oil revenues, with an increase in gross domestic product from $2,800 million in 1960 to $29,810 million in 1979 (World Bank 1981*b*, table 3). The size of its population, the potential for development of industry and agriculture, the availability of infrastructural facilities, and its relatively low oil reserves combine to make Algeria impatient with oil prices and revenues. The linkages among the sectors of its economy are investigated quantitatively later in this chapter.

## *Iran*

Before the revolution of 1979, the Iranian economy grew rapidly for almost two decades, with an average growth rate of 7.9 percent a year in GDP between 1960 and 1978.[3] A major transformation of the economy was achieved in the late sixties, for, in spite of increases in oil revenues, stagnation was evident through most of the fifties, and growth in agricultural output, in particular, lagged behind population growth. The biggest flaw in Iran's development policy was in its agricultural sector, which suffered from glaring neglect. The share of agriculture in gross domestic product declined from 29 percent in 1960 to 9 percent in 1978, and even though the growth rate for agriculture increased from 4.4 percent a year in 1960–70 to 5.2 percent a year in 1970–78, this contrasted with corresponding growth rates for manufacturing of 12.0 percent and 16.1 percent.[4] Neither agricultural nor industrial policy provided for exports. Fuel, minerals, and metals constituted 97 percent of the country's merchandise exports in 1960, increasing to 99 percent in 1977.

3. World Bank, *World Development Report 1980b*, table 1.
4. Ibid., tables 2, 3.

The industrial sector has witnessed substantial growth, but its linkages with the agricultural sector are weak—very little industrial output is used as inputs for agriculture. On the other hand, the intensive industrialization embarked upon under the former Shah raised the import coefficient of industry (with extensive imports of capital goods and machinery). During 1960–70, durable consumer goods constituted the largest proportion of industrial output. Industrial growth resulted in increased employment during the sixties but was accompanied by some negative effects. First, the industrial sector relied on massive imports of raw and semiprocessed materials, which ranked second only to the import bill for industrial machinery. Second, technology and capital equipment were imported almost indiscriminately from high-income countries without regard to technologies more suitable to Iran's low wage rates. As a result, the benefits of industrialization were severely restricted, and resources appear to have been misallocated.

Information on economic policies and developments since the revolution has been meager and unreliable; hence, an assessment of current trends is difficult. There is some indication, however, that there is an upswing in agricultural activity, insulated as rural areas have been (in relative terms) from the political turmoil since late 1978. Industry, on the other hand, has undoubtedly suffered a setback due to a paucity of foreign exchange because of both reduced oil revenues and the severance with the United States, once Iran's main source of technology and capital equipment. The slowdown in industry coupled with the enormous cost of the Iran-Iraq war is a strong justification for Iran trying to increase oil exports and revenues. But these economic compulsions will be tempered by political developments, which are unpredictable at present.

## *Iraq*

A recent article on Iraq describes it as the emerging power in the Middle East (Wright 1979). No doubt this scenario was influenced by the apparent collapse of Iran's power after the revolution and the progress in Iraq in the late seventies. The Iran-Iraq war may have thrown a damper on Iraq's emergence as a regional power, but that does not detract from its potential for economic progress. Of all the oil-exporting nations in the Middle East, Iraq is the only one with an abundance of water, cultivable land, mineral resources (besides petroleum), and a sizable but not too large population with relatively high levels of education. Despite these assets, little development took place in the fifties and sixties, largely because political events sapped society's productive efforts. Until the revolution of 1958, political and economic power was concentrated in the hands of the rich and privileged. But the revolution produced few gains, and the economy continued to stagnate, leading to the uprising of 1968, and bringing the Ba'athist regime of Al-Bakr and Saddam Hussein to power. In the years following this political change, development on all fronts was aided by increases in oil exports (until the start of the Iran-Iraq war) and oil prices. Saddam Hussein replaced Al-Bakr as the country's leader, but this change only ensured continuance of heavy-handed repression and total intolerance of political opposition. The government has followed a path of socialism, with a firm hold on economic decisions and resource allocation. Consequently, Iraq's oil policy is dictated by the party machinery, which also decides the government's developmental policies and plans.

Iraq's gross national product grew at an annual rate of 4.6 percent per capita in the period 1960–79. Industrial growth reached a level of 13.6 percent a year between 1960 and 1970, but in the same period, agricultural growth dropped from 5.7 percent a year to −1.8 percent a year. The country's dependence on exports of fuel, minerals, and metals increased from 97 percent in 1960 to 99 percent in 1978. As in the case of Iran, Iraq has done little to develop the full potential of its agriculture. Agrarian reforms have not yet made any dent on production; agricultural technology is influenced by traditional methods, which wastes water and does not rotate crops. Agriculture is also beset with inadequate marketing services and credit institutions, but irrigation and flood control measures are reasonably well developed.

The commendable feature of Iraq's development is the use of oil revenues: imports of consumer goods are severely regulated and 50 percent of oil revenues are allocated to development plans. But the highly centralized system and the repressive regime have bred extreme caution and inefficiency into an otherwise well-educated bureaucracy, which is the main instrument for planning and implementation of development. Undoubtedly, the lack of participation by society at large freezes Iraqi development into a mold of inefficiency, which does not do justice to the potential of this nation.

## *Kuwait*

Many have praised the near miraculous transformation of Kuwait physically, socially, and financially, from the immediate postwar years to today. Until the first shipment of oil was made in 1946, Kuwait was a typical developing society, with fishing, minor shipbuilding, and external trade as its main occupation. But the fishing industry was soon overtaken by technological change in other parts of the world, and its primary trading activity in cultured pearls was overtaken by Japan.

Oil revenues started contributing to economic development with the advent of the fifty-fifty profit-sharing arrangement in 1952. Also in 1952, the Kuwaiti government set up a development board; and in 1956 a general budget was introduced with the intention of organizing and expanding the financial resources of the economy. The increase in construction activities in the fifties attracted expatriate labor, to whom the government offered attractive terms in order to meet its labor needs. A high dependence on foreign labor exists even today. Kuwait's independence in 1961 ushered in a new era of planned development, with large investments in infrastructure. In 1979, Kuwait had the world's highest gross national product per capita, at $17,100. The distribution of gross domestic profit was 81 percent in industry and 19 percent in services for that year.[5] Its merchandise exports were broken down into 90 percent for fuels, minerals, and metals, 1 percent for other primary commodities, and the balance for manufactured goods (World Bank 1981b, tables 1, 2, 3, 9). There are also adequate roads, ports, airports, and public buildings, a well-developed communications system, and modern banking and financial services. Education is widespread and generally ahead of population growth. But there is little scope for increases in agriculture in this desert land, given the lack of water.

Kuwait has extended its industrial activities overseas in the form of joint ventures with other nations, and the Kuwait Petroleum Corporation, established in the sixties, has grown into a successful and progressive business enterprise. It plans to increase its refining capacity to 750,000 barrels a day by 1985, allowing it to refine 60 to 75 percent of its crude output (Fesharaki and Isaak 1981, 13). There is already in existence a petrochemicals industry that produces ammonia, and the country is due to move into polymers and aromatics as the base for manufacture of plastics. Kuwait's sound financial investments overseas (including the financial markets of other countries), a relatively well-developed infrastructure, and a growing ability to export products based on oil and natural gas instead of crude oil allows Kuwait to absorb changes in oil revenues to an even greater degree than Saudi Arabia. But the size of its production being considerably lower than that of Saudi Arabia, it has nowhere near the same power in influencing the world market through changes in production and export of oil. Kuwait is generally content to follow the lead of Saudi Arabia and OPEC as a whole in deciding oil exports and prices.

5. World Bank 1981b, tables 1, 2 and 9.

## Libya

A glimpse into the economic and political history of Libya explains why, unlike other capital-surplus nations of the Middle East, it is faltering in its development. Before independence in 1952, Libya bore the burden of Italian colonial rule, which retarded its economic growth and the development of its natural resources. About one-eighth of the agricultural land was in the hands of Italian settlers, and the rest of the land was too poor to increase productivity. Infrastructure, such as ports, transport, and communications, was designed to serve Italian interests, and, in any case, most of these facilities were destroyed during World War II. After the war Libya suffered repressive caretaker governments from Britain and France. At the time of independence, Libya was a poor country, with 90 percent illiteracy, a shaky political structure rife with inefficiency and corruption, and a notable lack of natural resources. Oil revenues, however, started making an impact in the fifties, and progress was achieved in education, health, transport, power generation, construction, and housing. Over 40 percent of oil revenues and overseas assistance were invested in development, of which agriculture received a sizable share. But weaknesses in existing institutions, shortages of skilled personnel, and inefficient government organizations hampered progress in various sectors of the economy.

The year 1969 was a major turning point in Libya's history. The monarchy was overthrown, and Qaddafi seized political control of the country. The effects of Qaddafi's militant postures on oil prices are discussed in chapter 4; these developments came in the wake of, and perhaps were inspired by, major new discoveries of oil and the increase in oil revenues. The new government also made efforts to promote agriculture, which had been neglected during the preoccupation with oil development. Actual investment in agriculture rose from 5.7 million dinars (US$19.27 million) in 1968 to 2.3 million dinars (US$71.96 million) in 1971-72. But with a rising share of oil income, the contribution of agriculture to national income remained constant at around 3 percent of gross domestic product in the early seventies. By 1979, this share had declined to 2 percent, with manufacturing accounting for 3 percent and services 25 percent, and the industrial sector, which is essentially the oil industry, contributing 73 percent (World Bank 1981*b*, table 3). But it is significant that agricultural output grew 11.8 percent a year and manufacturing 18.9 percent for the period 1970-79, and that the gross national product per capita was $8,170 (World Bank 1981*b*, tables 1, 2). Libya has accumulated a surplus in its balance of trade from year to year, even though in the period 1970-79 exports declined at an annual average rate of 6.5 percent and imports increased at 16.8 percent. At the end of 1979, Libya held gross international reserves of $7.6 billion (World Bank 1981*b*, table 15). The value of foreign assets is largely unknown, but one source placed them at 1475 million dinars in June 1978.[6]

Libya has followed a policy of restraint in oil production and exports. The highest output was achieved in 1973 at 796,423 thousand barrels and the lowest was recorded in 1981 at 417,150 thousand barrels. Existing capacity could permit an increase in output to over one billion barrels a year, but Libya has not attempted to reach this level—its absorptive capacity is low and surpluses invested overseas have generally yielded very low returns in real terms.

The inflow of expatriate labor has been regulated carefully, resulting in a lower proportion of foreign labor than in Kuwait and Saudi Arabia. During the midseventies, shortages of labor were felt all across the economy, and port congestion slowed down developments that depended on imports of raw materials and capital goods. Libya also suffers shortages of trained managers. Because of Qaddafi's radical "partners, not wage-earners" scheme, most private enterprise is run by people's committees. The oil industry and banks, however, continue to be managed professionally and are relatively efficient.

The stability of the present Libyan regime has often been called into question, which raises doubts about the future of the nation's oil policies. But should the present power structure remain unchanged, there appears to be little reason to expect major changes in the immediate future.

### Saudi Arabia

Saudi Arabia's economic history is characterized by marked fluctuations in the fifties and early sixties, influenced first by the financially lax King Saud and then by the austere and fiscally responsible Crown Prince Faisal, who was appointed prime minister in 1958. Faisal was concerned about long-term economic change and set up the Committee for Economic Development, which later requested the World Bank to send a team to study Saudi Arabia's economic problems. The World Bank's report on the subject advocated increased allocations for development. Estimates of the kingdom's national income include those by the American University in Beirut and by the United Nations Economic and Social Office of Beirut.

The economy of Saudi Arabia grew at an annual rate of 12 percent between 1962 and 1973. This increase was attributable to the growth of

6. "Arab Banking and Finance," *Times* (London), March 11, 1980.

oil exports, which contributed an average of 52.6 percent of the gross domestic product between 1954–55 and 1964–65. By the end of the seventies, the share of industry as a whole had increased to 74 percent. Agriculture contributed only 1 percent, but its importance lies in the fact that over half the rural population is in some way connected with agriculture. Consequently, irrigation, drainage, and land reclamation received large investments during the seventies. But the impact of these investments like those in the industrial sector has been limited by shortages of trained labor and entrepreneurial talent, high costs of imported equipment and raw materials, insufficient infrastructure such as roads and communications, and inadequate comprehensive preinvestment studies. The problem of poor roads and communications is of special relevance to Saudi Arabia, on account of its vast area and sparse population. A massive expansion of the nonoil sectors of the Saudi economy has been undertaken, particularly since 1973. Gross domestic investment in a range of projects, from education to industry, grew at an annual rate of 46.7 percent during the period 1970–79 (World Bank 1981 *b*, table 4). But labor shortages, port congestion, and lack of expert management has caused considerable delay in construction projects.

Saudi Arabia's oil policies in the future are likely to be motivated by its desire for (1) continued domination of the world oil market, (2) continued political influence with its Arab neighbors, (3) harmonious relationships with the West and a pronounced anticommunist identity, (4) higher levels of consumption for its population, and (5) diversification of its production base. These factors have to be seen in relation to the conservative nature of the Saudi government and its vested interests in perpetuating a social structure dependent on constantly growing prosperity. Even though stopping oil exports and relying on income from foreign investments is theoretically feasible, it could not be sustained in practice. The Saudi economy has expanded to a point where oil revenues are essential for maintaining consumption levels and the momentum of developmental activities. A 50 percent reduction in these revenues could create problems of differential income growth. The social structure, which is cemented both by Islamic orthodoxy and rapid income increases for even the poorest, could be faced with a major upheaval. Yet the gap between the minimum desired production of five million barrels a day and the maximum sustainable capacity of twelve million barrels a day is enough to permit large discretionary swings in production in the future.

## United Arab Emirates, Bahrain, Qatar, and Oman

Pearl fishing has historically been the most important industry in the United Arab Emirates, but it declined in the fifties partly because of competition from Japan and partly because of the growth of the oil industry, which provided employment opportunities and alternative sources of income. The 1977 gross domestic product per person was estimated at $14,420.[7] Traditionally this region was known for its trade, particularly with India, and the oil boom has once again led to a great deal of trading, though mainly in reverse. The emirates have, since 1973, been on a crash program of economic development, importing large quantities of goods and services. The post-1973 boom also generated major changes in population. An influx of largely South Asian workers has created problems for the governments of these countries, which, while dependent on these workers, are at the same time increasingly apprehensive about the consequences to their countries' demographics. Despite the growth of the work force, the emirates remain low-absorbing economic systems. The emirate of Abu Dhabi is the primary supplier of funds for the federal budget of the UAE and also provides the major portion of UAE aid to other countries. Abu Dhabi was also the first emirate to implement a formal economic development plan. The institutions and processes for economic planning in the other emirates are not as sophisticated. Thus, even with political integration, economic policies among the emirates are not closely coordinated, and economies of scale are not fully realized because in economic matters national interests and concerns dominate decision making. The UAE, in addition to its oil-related industrial projects, manufactures construction materials, industrial gases, furniture, and printing presses. Most major industries, such as the Al-Nar oil refinery and the Al-Ain cement factory, are located in Abu Dhabi, but Dubai recently launched a major industrialization effort involving electricity generation, desalination, a dry dock, and gas-based industries. Ras Al Khaima, with its fertile soil, underground water, and relatively high rainfall, is the most advanced in agriculture and is making substantial efforts to introduce modern agricultural techniques by setting up experimental farms and training institutions. Economic integration is of considerable significance for efficient resource allocation in the UAE. The economic development of each emirate should be planned in the context of and as a component of the economic development of the federation.

Both Bahrain and Qatar, like the United Arab Emirates, trace their historical economic activities to pearl fishing, which in these countries

7. World Bank, *World Development Report 1979* (Washington, D.C.: 1979), table 1.

also waned during the fifties due to competition from Japan. Bahrain has been politically stable in the past, but recent attempts by a small faction to topple the government have been a source of concern to all Persian Gulf emirates. Bahrain has been a leader in banking activities in the region, with the first offshore banking venture among these countries. Oil in Qatar was found in 1939, but control of production was locally acquired in 1979, when the government nationalized the Qatar Petroleum Company. With a population of around a quarter million, and 1980 oil revenues over $5 billion, Qatar has generated surpluses beyond its capacity to absorb. The management of its financial reserves is entrusted to an investment board, and projects invested in locally include a 1.2-million-riyal (US $0.33 million) steel mill, a petrochemicals project, oil refineries, and gas liquefaction plants.

Allocation of resources and disposition of oil revenues follow similar patterns in most of the United Arab Emirates, Bahrain, and Qatar: large shares go to the social consumption sector, nonproductive ventures like public administration and armed forces, and transfer payments and outward remittances by the large migrant labor population. In addition, distribution of income is heavily skewed, and the ruling families and richer sheikhs maintain high consumption levels financed directly or indirectly from export of oil.

The sultanate of Oman stands separate from the other Persian Gulf states on account of its special political and military relationship with the United States. Oman also has a shorter history of oil production than the others, with its first major oil field having been found, at Fahrid, only in 1974. Earlier, Oman was heavily dependent on external assistance and until 1975 incurred large military expenditures fighting a civil war in Dhofar. Oman received Arab funds in 1976 for a pipeline to bring gas to the coast for industrial projects. A projected refinery at Muscat and other industrial projects are likely to bring about a rapid expansion in Oman's economy in the next five to ten years.

The analysis of these countries has been necessarily brief and purely descriptive. But the historic and political realities of their economic growth underlie their decisions on oil production and export. Their discretionary power to increase or decrease oil production, quite apart from technical constraints, is intimately connected with their perception of the effects of changes in oil revenues on their economic and political strengths.

At the same time, these countries are undergoing a major transformation from traditional (and even primitive) societies to modern states, and a quantitative assessment of their economic growth patterns would be useful in our understanding of the subject. Most of these countries are precariously dependent on their oil sectors and realize that they have to use this geostructurally transient boon to diversify their economies and reduce their dependence on oil. Investments in diversification have, therefore, greatly changed their patterns of growth. I have subjected these patterns to econometric analysis, using simple macroeconomic national income models for selected countries of the region (Saudi Arabia, Libya, Kuwait, Iran, Iraq, and Algeria; the UAE, Bahrain, Qatar, and Oman had to be excluded due to inadequacy of data).

The models are based on the structural characteristics exhibited by each nation's economy in recent years and are, therefore, to be used with extreme caution in making long-run predictions. Specification is based on the composition of output in the countries modeled. Each model has been disaggregated to provide a suitable explanation of a country's economic output. The value-added measures in the major sectors, in the form of sectoral contributions to gross domestic product, are the building blocks of the models.

In specifying the models, I was guided by recent high rate of investment in economic infrastructure. Most oil producing countries realize that any shift from an oil-dependent economic structure is possible only if infrastructural facilities are created in the medium term. Investment in infrastructure is, therefore, expected to play a significant role in contribution to gross domestic product in the remaining years of this century. Other sectors considered are agriculture, industry, wholesale and retail trade, and of course transport and communications, which can be considered a component of infrastructural facilities. Of the countries included in this part of the study, only Iran and Iraq have significant contributions from the traditional agricultural sector. Notably, these two countries have not suffered from colonialism. In general, the nonoil component of the industrial sector has not contributed to gross domestic product to any significant degree, because efforts at industrialization are very recent and industrial activity, except for oil and related production, has not reached maturity.

To allow for the foreign trade sector in the models, I constructed a purchasing power parity index to measure the excess purchasing power in the international market, which accrues to each country for import of goods and services in relation to the export price of petroleum. The components of this index are (1) the export price index for petroleum in each country, $P_p$; (2) United Nations index of export price of manufactured goods, $P_m$; and (3) the balance-of-trade statistics for each country, $B_T$.

The index is then calculated as $\dfrac{P_p - P_m}{P_m}(B_T)$.

Other variables are

$Y$ (gross domestic product), $X_1$ (share of agricultural sector), $X_2$ (share of industrial sector), $X_3$ (share of construction sector), $X_4$ (share of wholesale and retail trade), $X_5$ (share of transport and communications), $X_6$ (share of mining and quarrying, including crude oil), $X_7$ (share of other sectors), $X_8$ (net export earnings), $X_9$ (increase in capital stock), $X_{10}$ (supply of money), $X_{11}$ (purchasing power parity index), $X_{12}$ (a dummy variable representing the specific country). In its most general form, the model specified was

$$Y = f(X_1, X_2, \ldots, X_{12})$$

$$\text{and} \quad \frac{dY}{dX_i} > 0.$$

In order to estimate the model I pooled time-series and cross-section data for homogeneous groups of countries, since adequate time series were not available for any of the countries included in our study. The groupings for which models were estimated include Algeria, Iran, Iraq; Saudi Arabia and Libya; and Saudi Arabia, Libya, and Kuwait. These groupings were ordered on the basis of similarities of structure and absorptive capacity between the countries. Sources of data for estimation included the *United Nations Year Book of National Income Accounts, Balance of Payments Statistics, United Nations Statistical Year Book*, and the *Yearbook of International Financial Statistics* published by the International Monetary Fund. Since data were available only in current prices, the United Nations export price index of manufactured commodities was used where appropriate as a price deflator. With the exception of net export earnings, which were measured in special drawing rights, all other variables were initially measured in the national currency of each country and converted to U.S. dollars for the regression runs.

Some estimates resulting from this exercise are shown in table 5.3. These were obtained by performing step-wise regression, using ordinary least squares procedure with data for the period 1967–75—hence the results show trends somewhat longer than the boom conditions immediately following 1973. The results shown in table 5.3 indicate that, based on the linear form of the model, low absorbers Saudi Arabia and Libya are dependent largely on construction, wholesale and retail trade, mining and quarrying, and "other sectors". But in the log linear form, the construction sector and purchasing power variable turn out to be the most significant. When Kuwait is added to the grouping, however, the construction sector turns out to be the most dominant in all three countries. The importance of the purchasing-power variable rests on the

TABLE 5.3

Econometric Estimates of the Composition of Gross Domestic Product, Selected OPEC Countries (billions of dollars)

| Country | Form | Dependent Variable | Constant | $X_1$ | $X_3$ | $X_4$ | $X_6$ | $X_7$ | $X_{11}$ | $R^2$ |
|---|---|---|---|---|---|---|---|---|---|---|
| *High absorbers* | | | | | | | | | | |
| Algeria | Linear | GDP | 6.9 | 2.20 (4.61) | — | 8.90 (6.14) | 0.840 (13.43) | — | — | 0.99 |
| Iran | Linear | GDP | −3.3 | 2.20 (4.61) | — | 8.90 (6.14) | 0.840 (13.43) | — | — | 0.99 |
| Iraq | Linear | GDP | −60.5 | 2.20 (4.61) | — | 8.90 (6.14) | 0.840 (13.43) | — | — | 0.99 |
| Algeria | Log Linear | GDP | −6.64 | 2.19 (4.45) | — | 8.67 (5.91) | 0.084 (13.50) | — | — | 0.99 |
| Iran | Log Linear | GDP | −2.54 | 2.19 (4.45) | — | 8.67 (5.91) | 0.084 (13.50) | — | — | 0.99 |
| Iraq | Log Linear | GDP | −57.14 | 2.19 (4.45) | — | 8.67 (5.91) | 0.084 (13.50) | — | — | 0.99 |
| *Low Absorbers* | | | | | | | | | | |
| Saudi Arabia | Linear | GDP | −37.47 | — | 0.70 (2.73) | 3.96 (3.96) | 1.050 (222.76) | 1.19 (3.75) | — | 0.99 |
| Libya | Linear | GDP | −1.37 | — | 0.70 (2.73) | 3.96 (3.96) | 1.050 (222.76) | 1.19 (3.75) | — | 0.99 |
| Saudi Arabia | Linear | GDP | 128.59 | — | 3.47 (7.48) | — | 1.050 (24.49) | — | 0.0012 (1.07) | 0.99 |
| Libya | Linear | GDP | 36.96 | — | 3.47 (7.48) | — | 1.050 (24.49) | — | 0.0012 (1.07) | 0.99 |
| Kuwait | Linear | GDP | 68.15 | — | 3.47 (7.48) | — | 1.050 (24.49) | — | 0.0012 (1.07) | 0.99 |

[a] $X_1$, agricultural; $X_3$, construction; $X_4$, wholesale and retail trade; $X_6$, mining and quarrying (including crude oil); $X_7$, other sectors; $X_{11}$, purchasing power index. Figures in parentheses are *t*-ratios.

fact that for these countries, being low absorbers, external purchase of goods and services is a better determinant of domestic income than the value of oil production.

The estimates shown in table 5.3 are to be used for a limited purpose only and can at best help in cross-checking results from other methods. The coefficients of the independent variables represent the multiplier effect of each sector on the particular country's economy. Consequently, given the rate of growth of each major sector, it is possible to arrive at forecasts of gross domestic product growth in the future. Sectoral growth rates in turn can be predicted on the basis of a specific assessment of each country's economy and the absorptive capacities of its major growth sectors. Because of rapid structural changes in these countries, prediction of growth of specific activities and the shares of different sectors in each country's economy is difficult. Consequently, the absorptive capacity and therefore the effect of oil revenues for each country is also difficult to estimate.

Estimates of growth were developed as follows: First, scenarios of sectoral growth rates were developed for Iraq, Kuwait, Libya, Qatar, Saudi Arabia, and the United Arab Emirates. Their growth of gross domestic product (except for Qatar and the UAE) was estimated within the framework of the income-accounting models described in table 5.3. Second, it was assumed that these countries would face no shortage of foreign exchange, and the sectoral growth rates would not be constrained by limits in external resources. Third, scenarios of growth as developed above (sectoral as well as gross domestic product) were communicated to a group of twelve knowledgeable persons—academics, senior executives of organizations doing business in these countries, and researchers in the Arab world. Fourth, on the basis of eight responses, I developed two sets of consensus scenarios of growth of nonoil gross domestic product. A majority of the respondents painted very general scenarios of growth in output of oil, and I, therefore, forecast only nonoil gross domestic product.[8]

The growth scenarios thus arrived at are shown in table 5.4. These predictions rest on projections of slowing absorptive capacities in these countries and deliberate policies of cooling down development. Consequently, rates of growth are seen as tapering down toward the end of this century, with implications for investments and absorptive capacity that I will assess later. An important, though unavoidable, omission in these projections is that of Iran. But any forecast of economic trends and policies in Iran is complicated by the uncertainties of both its

---

8. The respondents were informed that their opinions would not be quoted and that responsibility for the consensus forecasts would rest with me.

TABLE 5.4

Projected Rates of Annual Growth of Nonoil Gross Domestic Product, Selected OPEC Countries, 1982–2000 (percent)

| Country | 1980–82 | 1982–85 | 1985–90 | 1990–2000 |
|---|---|---|---|---|
| | | Low-growth scenario | | |
| Iraq | 20 | 15 | 10 | 8 |
| Kuwait | 19 | 15 | 9 | 6 |
| Libya | 15 | 12 | 8 | 6 |
| Qatar | 21 | 16 | 12 | 7 |
| Saudi Arabia | 25 | 16 | 12 | 8 |
| United Arab Emirates | 20 | 15 | 10 | 7 |
| | | High-growth scenario | | |
| Iraq | 20 | 18 | 17 | 15 |
| Kuwait | 19 | 15 | 12 | 9 |
| Libya | 15 | 15 | 10 | 8 |
| Qatar | 21 | 20 | 15 | 10 |
| Saudi Arabia | 25 | 25 | 20 | 15 |
| United Arab Emirates | 20 | 18 | 12 | 10 |

political future and the extent to which its economy will open to the outside world. Since Iran is by far the most powerful nation in the Persian Gulf (as its endurance in the Iran-Iraq war amply establishes), it will no doubt adopt a more dominant posture in the affairs of the region in the future. But how and when it will emerge from its isolation is difficult to predict. For Qatar and the UAE, I had no quantitative framework within which to evaluate sectoral growth rates; the respondents were asked to estimate nonoil growth in gross domestic product based on past trends and existing information. The scenarios for these countries, therefore, also reflect two consensus views.

Nonoil gross domestic product cannot be separated from the oil sector, and in fact the purpose of our discussion in this chapter is to establish the importance of development goals, plans, and achievements as a determinant of oil output, price, and exports. But merely looking at oil revenues and their domestic absorption does not provide us with enough information for realistic assessments of probable changes in oil-related variables. While the link between domestic absorption (economic development) and oil revenues is a dominant factor in the future of oil prices and exports, a complex set of global and interregional influences are also of major importance in this area. The international dimensions of this subject are discussed in the following chapters.

CHAPTER 6
# OPEC AND THE GLOBAL ECONOMY

The paramountcy of OPEC in the global oil market was established in 1973, but initial assessments of its global impact ranged from calamitous to a mere flash in the pan. More than ten years after the oil embargo of 1973, there are still conflicting analyses on the extent of OPEC's power. Initial apprehensions included a massive transfer of resources from major oil-consuming nations of the West to OPEC and stagnation or economic crashes in countries of the Third World with high oil-import dependence. It was also feared that unprecedented petrodollar flows would cause a crisis in the international monetary system, as it switched from the assets from one currency to another. While the global economic system appears to have weathered the storm, the effects of OPEC's decisions have brought about many changes of a lasting nature.

OPEC occupies the apex of a triangle of which the other two points are the developed nations of the North and the developing nations of the South. OPEC has consistently drawn the support of the South, since the changes it has wrought are seen as forerunners of changes the Third World has been demanding for decades in its quest for a new international economic order. The successful cartelization of oil producers was seen as a model for similar structures for commodities exported by the Third World. But their support of OPEC was painful and embarrassing to them in view of their own higher oil-import bills and the direction of petrodollar flows toward the countries of the North.

But, in relative terms, higher oil prices have also had an unfavorable impact on the economic output and welfare of the countries of the North, grown accustomed as they were to declining real prices of oil and increasing dependence on oil use. That they have been able to minimize and contain the economic ill effects of more expensive energy is a tribute to their resilience and the strengths of their institutions. The oil shock of 1979–80 was in many ways different from that of 1973–74. Inflation in the industrial world was much lower than in the wake of the 1973–74 oil crisis. According to the Bank for International Settlements,

TABLE 6.1

Changes in Consumer Prices, Industrial Countries, 1973–81 (percent)

| Country | 1973[a] | 1974[a] | 1975[a] | 1976[a] | 1977[a] | 1978[a] | 1979[a] | 1980[a] | 1981 |
|---|---|---|---|---|---|---|---|---|---|
| United States | 8.8 | 12.2 | 7.0 | 4.8 | 6.8 | 9.0 | 13.3 | 12.4 | 10.0[b] |
| Japan | 19.0 | 22.0 | 7.7 | 10.4 | 4.8 | 3.5 | 5.8 | 7.1 | 6.2[c] |
| United Kingdom | 10.6 | 19.2 | 24.9 | 15.1 | 12.1 | 8.4 | 17.2 | 15.1 | 12.0[b] |
| Italy | 12.3 | 25.3 | 11.1 | 21.8 | 14.9 | 11.9 | 19.8 | 21.1 | 19.9[b] |
| Canada | 9.1 | 12.5 | 9.5 | 5.8 | 9.5 | 8.4 | 9.8 | 11.2 | 12.6[b] |
| France | 8.5 | 15.2 | 9.6 | 9.9 | 9.0 | 9.7 | 11.8 | 13.6 | 12.5[c] |
| Sweden | 7.5 | 11.6 | 8.9 | 9.6 | 12.7 | 7.4 | 9.8 | 14.1 | 12.9[b] |
| West Germany | 7.8 | 5.8 | 5.4 | 3.7 | 3.5 | 2.5 | 5.4 | 5.5 | 5.6[b] |
| Switzerland | 11.9 | 7.6 | 3.4 | 1.3 | 1.1 | 0.7 | 5.1 | 4.4 | 5.6[b] |
| Belgium | 7.3 | 15.7 | 11.0 | 7.6 | 6.3 | 3.9 | 5.1 | 7.5 | 7.4[b] |
| Netherlands | 8.2 | 10.9 | 9.1 | 8.5 | 5.2 | 3.9 | 4.8 | 6.7 | 6.2[b] |
| Austria | 7.8 | 9.7 | 6.7 | 7.2 | 4.2 | 3.7 | 4.7 | 6.7 | 7.4[b] |
| Denmark | 12.6 | 15.5 | 4.3 | 13.1 | 12.2 | 7.1 | 11.8 | 10.9 | 11.3[c] |
| Finland | 14.1 | 16.9 | 18.1 | 12.3 | 11.9 | 6.5 | 8.6 | 13.8 | 13.1[c] |
| Greece | 30.7 | 13.5 | 15.7 | 11.7 | 12.8 | 11.5 | 24.8 | 26.2 | 24.3[b] |
| Norway | 7.6 | 10.5 | 11.0 | 8.0 | 9.1 | 8.1 | 4.7 | 13.7 | 14.6[b] |
| Ireland | 12.6 | 20.0 | 16.8 | 20.6 | 10.8 | 7.9 | 16.0 | 18.2 | 21.0[d] |
| Spain | 14.3 | 17.9 | 14.1 | 19.8 | 26.4 | 16.6 | 15.5 | 15.1 | 13.8[d] |

[a] January 1 to December 31.
[b] January 1 to April 30.
[c] January 1 to March 31.
[d] January 1 to February 28.

consumer prices in the Group of Ten nations and Switzerland rose by about 24 percent during the first period and 10.5 percent during the second.[1] The pattern of price rises was also different, with a much slower and smoother acceleration in the second period than in the first. It is generally accepted that there is a strong link between upward ratcheting of oil and commodity prices and the consumer price index in industrial countries, but the precise extent of this inflationary effect varies from country to country, as shown in table 6.1.

Among the industrial nations some broad groupings can be identified. In the United States, Canada, France, Norway, and Sweden, inflation rates in 1973 and 1978 respectively, were more or less the same, that is, between 7.5 and 9.7 percent. After a steep climb, inflation rates came down to between 7.0 and 11.0 percent two years after the first shock. However, they remained at much higher levels at a corresponding time after the second shock. Inflation in the United Kingdom, Italy, and Spain was also at similar levels at the time of the two oil shocks, but much higher than the first group. The United Kingdom held down

1. *Fifty-first Annual Report* (Basle: 1981), p. 9.

inflation after the second shock, assisted by its own substantial production of oil, while Italy did much worse in this regard. Japan, West Germany, Belgium, the Netherlands, Austria, and Switzerland halved inflation rates between 1973 and 1975 (from 7.0–19.0 percent to 3.4–11.0 percent). But between 1978 and 1980, inflation increased from 0.7–3.9 percent to 4.4–7.5 percent.

These differences among groups and within groups were inevitable, since significant differences existed in dependence on oil imports, in exchange rates, and in responses to inflation. But the feature common to all nations is the speed at which the second price shock passed through the system (in contrast with the first shock, when most policymakers appeared reluctant to raise energy prices for the consumer). The U.S. government removed price controls on domestically produced oil, resulting in a rapid increase in gasoline prices. This and similar actions in other countries had a pronounced effect on energy consumption. Of course, demand lags behind increases in price, so much of what is happening in the eighties is the result of price increases in the late seventies.

The two periods (1973–74 and 1970–80) were also marked by increases in prices of other commodities. The second-round increases were milder for various reasons. First, commodity prices in general have followed a downward trend over the past decade; temporary increases were soon followed by decreases, often larger. Second, following OPEC's price increase of 1973–74, most producers of commodities felt encouraged to increase prices, and some countries made serious efforts to cartelize, encouraging speculative buying and a further upward push in prices. Third, the commodity market in mid-1973 was very strong in contrast to early 1979. Changes in oil and nonoil commodity prices from 1972 to mid-1981 are shown in figure 6.1. Whereas oil prices maintained an upward trend through the seventies, prices of nonoil commodities continued their historic pattern of fluctuations, and in real terms have gone down, since nominal prices have remained more or less constant. These fluctuations reflect both changes in business activity in the industrial nations and changes in supply, particularly of agricultural commodities such as sugar, coffee, and food grains. Crop shortages and failures result in sudden price increases, which are amplified by speculative buying, and these generally last at least one full agricultural cycle, or until a good crop increases supplies. Prices then fall until the same phenomenon is repeated.

The link between energy prices and economic activity in the industrial nations is not conclusive, and research results vary. Following the 1973–74 price increase, there was a short-run net loss in aggregate demand as shown in table 6.2. This method of simple accounting, of

OPEC and the Global Economy / 97

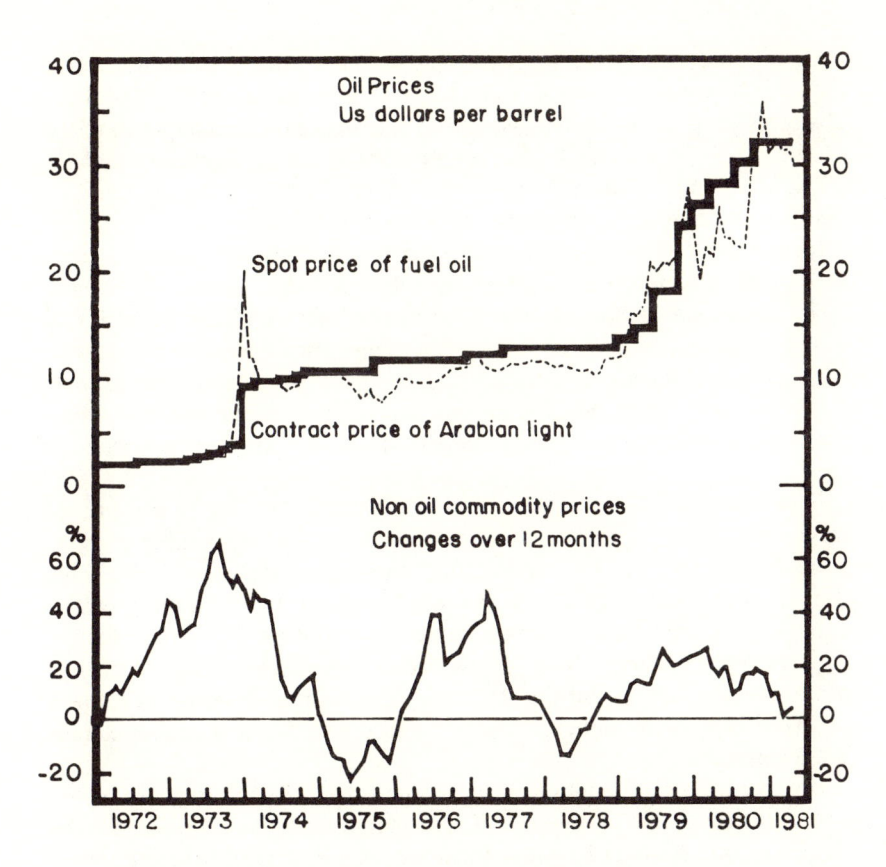

Figure 6.1 World-Market Commodity Prices; Oil and Nonoil
*Source*: Bank for International Settlements, *Fifty-first Annual Report* (Basle: 1981).

TABLE 6.2

Changes in Components of Demand, Industrial Countries, 1974
(percent of gross domestic product)

| Country or Region | Increase in Oil Import Bill | Increase in Exports to Oil-Exporting Countries | Increase in Payments to Domestic Oil Producers | Loss of Exports to Oil-Importing Developing Nations | Total |
|---|---|---|---|---|---|
| United States | −1.03 | +0.18 | −0.86 | −0.03 | −1.74 |
| Western Europe | −2.28 | +0.35 | — | −0.03 | −1.96 |
| Japan | −2.67 | +0.49 | — | −0.06 | −2.24 |

*Source*: Edward R. Fried and Charles L. Schultze, eds., *Higher Oil Prices and the World Economy* (Washington, D.C.: Brookings, 1975), table 1–7.

course, reviews only the short-run impact, and it does not include other global influences and factors that may have determined any part of these changes. It is, nevertheless, useful for evaluating these changes in terms of the 1973–74 oil shock. The implication of these figures is that, in 1974, aggregate demand as a percentage of gross domestic product declined by similar amounts in the three country groups. The long-run effects of the 1973–74 price increase were, of course, very different from those indicated in table 6.2. With the lull in oil prices between 1975 and 1978, there was a decline in the oil import bill for these countries, a massive surge in exports to the oil-exporting nations, a minor increase in payments to domestic oil producers, and steady improvement in exports to the oil-importing developing countries. The adjustment was facilitated by the recycling of petrodollars from the oil-exporting countries to the industrial countries.

The fears in 1973 about the effects of oil price increases, in the form of large investible surpluses, flooding the international banking system, were belied for reasons which merit discussion. Balance-of-payments deficits have occurred since the beginning of international trading activities; if any country runs a current-account deficit while another runs a corresponding surplus, funds have to be recycled from the surplus nation to the deficit nation. This simple adjustment would become a problem only if the deficit is chronic and the underlying causes cannot be removed.

An example of a solution to a chronic surplus-deficit problem is the Marshall Plan, which helped recycle large U.S. surpluses into the deficit-ridden postwar nations of Europe. Again in the fifties and sixties the surpluses of the industrial nations were recycled into the developing contries, through lending programs and official development assistance. The OPEC surplus, even though successfully handled, has some features different from the two experiences mentioned above. The first of these is the emergence of OPEC as a major exporter, with recurring surpluses over long periods. The second is the large 1980 deficit of the countries of the Organisation for Economic Cooperation and Development. Other dimensions of the recycling process are shown in table 6.3. It is useful to observe that despite a large increase in surpluses and deficits since 1973, the total volume of recycling was still under 5 percent of the gross domestic product of the OECD nations. But while the OECD nations have adjusted remarkably well, the developing countries have been much slower in reducing their deficits. The reason for this lies in the fact that, while the industrial nations need working capital only at a time of deficits, developing countries need capital for long durations for restructuring their production base. The international banking

TABLE 6.3

Recycling of International Deficits and Surpluses, Country Groups, Selected Periods, 1950–80 (billions of dollars)

| Country Group and Item | 1950s | 1960s | 1973 | 1978 | 1974–79 | 1980 |
|---|---|---|---|---|---|---|
| | | | Current prices | | | |
| OPEC export revenues | 6 | 10 | 39 | 141 | 122 | 294 |
| OPEC balance | 0 | 1 | 9 | 9 | 40 | 126 |
| OECD balance | 3 | 3 | 18 | 28 | 0 | −44 |
| Oil-importing developing countries' balance | −3 | −7 | −12 | −48 | −43 | −82 |
| Rest of noncommunist world balance | 0 | 3 | −15 | 11 | 3 | 0 |
| OECD GDP | — | 1520 | 3247 | 5909 | 5069 | 6740 |
| Total deficit (surplus) | — | 7 | 27 | 48 | 43 | 125 |
| | | | 1978 prices | | | |
| OPEC balance | 0 | 3 | 17 | 9 | 44 | 91 |
| OECD balance | 9 | 9 | 33 | 28 | 0 | −44 |
| Oil-importing developing countries' balance | −9 | −21 | −22 | −48 | −48 | −47 |
| Rest of noncommunist world balance | 0 | 9 | −28 | 11 | 4 | 0 |
| OECD GDP | — | 3533 | 5228 | 5909 | 5616 | 5968 |
| Total deficit (surplus) | — | 21 | 50 | 48 | 48 | 91 |

*Source*: J. Nicholas Robinson, "The Role of Oil Funds Recycling in International Payments and Adjustment Problems," *OPEC Review* 4(2), table 1; and International Monetary Fund, *World Economic Outlook* (Washington, D.C.: 1981), table 14.

system, working on commercial principles which include the evaluation of risk, has not been as effective as, say, the Marshall Plan in restructuring these economies to eliminate the cause of continuing deficits.

A major observation on inflation in OECD countries can be made by looking at the price changes shown in figure 6.1 and the deficits shown in table 6.3. Because of passive policies, these nations experienced high rates of inflation during the period 1973–78, even though oil prices remained relatively stable after 1975. In effect, therefore, the value of exports from OECD nations to OPEC nations went up sharply, as much due to increased volume for development in OPEC nations as to inflated prices of goods in OECD nations. A similar adjustment took place in some of the developing countries as well, but its magnitude and speed were considerably lower. The lessons from the petrodollar recycling of the 1970s can, therefore, be summed up as follows:

1. Most OPEC investors placed their surplus funds in the Eurodollar market, essentially shifting the balance of surplus holdings in favor of the OECD nations. These nations in turn loaned large amounts to the developing countries, using the institutional strengths of their banking systems.
2. OECD countries borrowed large amounts, bringing about a rapid adjustment of their economies; this adjustment was aided by lower real prices of oil, domestic inflation, and large imports of their goods and services by OPEC nations.
3. The initial preference of OPEC investors was for short-term deposits, but between 1973 and 1978 the terms of maturities lengthened significantly.
4. Developing countries were unable to adjust to increased oil prices. Some of them borrowed heavily from the European banking system, but they were generally not able to change their structures and patterns of production to overcome their burdens of debt.
5. Investors earned very low rates of return, and in certain periods the returns were actually negative in real terms.

It would be useful to analyze the flows from the OPEC countries to the European banking system, and the motives that inspired the

TABLE 6.4

Deployment of OPEC Investible Surplus, 1974–79 (billions of dollars)

| Investment | 1974 | 1975 | 1976 | 1977 | 1978 | 1979 | Level End of 1979 |
|---|---|---|---|---|---|---|---|
| Bank deposits, country of currency issue | 7.0 | 2.0 | 0.5 | 2.3 | 2.8 | 6.3 | 26 |
| Bank deposits, Eurocurrency | 21.6 | 7.9 | 11.5 | 10.7 | 1.1 | 31.0 | 89 |
| Short-term government securities | 8.0 | −0.4 | −2.2 | −1.1 | −0.8 | 3.3 | 7 |
| Long-term government securities | 1.1 | 2.4 | 4.4 | 4.5 | −1.8 | −0.7 | 10 |
| Other capital flows | 7.1 | 12.8 | 13.2 | 9.8 | 5.8 | 9.0 | 58 |
| I.M.F. and I.B.R.D. | 3.5 | 4.0 | 2.0 | 0.3 | 0.1 | 2.0 | 46 |
| Developing countries | 4.9 | 6.5 | 6.4 | 7.0 | 6.2 | 6.9 | 236 |
| Total | 53.2 | 35.2 | 35.8 | 33.5 | 13.4 | 53.8 | * |
| Unidentified items | 1.9 | 1.1 | 2.8 | 4.1 | 5.4 | 25.2 | * |
| Total cash surplus | 55.1 | 36.3 | 38.6 | 37.6 | 18.8 | 79.0 | * |

Source: Bank of England Quarterly (June, 1980); estimated from recipient country information.
* Not known.

investments. The deployment of OPEC's investible surplus during 1974–79 is shown in table 6.4. In 1974, there was an immediate high concentration in bank deposits and short-term government securities. In the subsequent three years, the trend was in favor of long-term government securities and other investments. But, despite impressions to the contrary, funds provided for direct investments abroad, including holdings of property by individuals, were meager. In the first-round recycling process, OPEC foreign assets exhibited a strong regional concentration, with more than a quarter of all assets (27 percent) invested in the United States and 30 percent in the United Kingdom. (This percentage includes Euromarket deposits, 50 percent of which were made in the United Kingdom.) But investments began diversifying in 1976, with larger amounts flowing to Germany, Switzerland, and Japan.[2]

To understand OPEC investor behavior, it is important to remember that over 75 percent of OPEC's net foreign assets are held by Saudi Arabia, Kuwait, Libya, the United Arab Emirates, and Qatar. These nations, though transformed very rapidly into global economic powers, are deeply rooted in tradition and are, in some ways, developing nations. They are for political and social reasons insecure about the future, and even though the expected lives of their known oil reserves at current production levels are over forty years (and much higher in the case of Saudi Arabia and Kuwait), they envision an oilless era. Current-account surpluses are seen as transient and would provide only low returns if ploughed back into their own economies. OPEC investors' cautious behavior is influenced heavily by the premium that they attach to security and by the fact that OPEC is a new entrant in the international financial markets. OPEC has provided generous aid to the developing nations for reasons of Islamic brotherhood and of fostering a Robin Hood image. But it has shunned direct investment in them because of the risks. In sum, OPEC investors' highest priority is safety; high financial returns are welcome but not essential. They prefer investing in private institutions but do not shun investments in government securities and bonds. Serious currency losses have influenced them to diversify among currencies, but in doing so they have exercised caution, aware that a movement away from the dollar might cause a run on this currency in international money markets and jeopardize both the value of their assets and the real price of oil. There

---

2. In fifteen months after the 1973 oil crisis, sellers of Japanese shares off-loaded $1.5 billion worth of stock, but by 1979, $5.97 billion worth Japanese shares were bought in about nine months, which brokers indicate involved Middle East investors. See James Bartholomew, "Tokyo's Secret Petro-dollar Connection," *Far Eastern Economic Review*, July 1980, p. 38.

are of course significant differences in financial behavior among members of OPEC. The Kuwaitis invest in blue scrips. From the information they are required to divulge in Great Britain, it can be inferred that they are enterprising and businesslike equity investors willing to take risks. The Saudis, on the other hand, prefer stable fixed investments and are averse to risk-taking.

Insights into OPEC investor behavior can also be gleaned from returns from foreign investments. Unfortunately, data are not available to provide a comprehensive picture, but there is enough evidence to establish that financial returns have generally been low. Table 6.5 shows gross rates of return for the period 1976-78. During this period the length of maturity of assets was increasing, and some diversification of assets was also taking place, bringing higher rates of return at the end of the period than at the beginning. But considering that this was a period of high inflation in the OECD countries, these returns are very low and actually imply an erosion in the value of OPEC's foreign assets.

The importance of the petrodollar recycling process lies not only in its effects on international finance and the adjustment of other countries to higher oil prices but also in its implications for OPEC's posture on oil prices. By 1977 and 1978, OPEC had started clamoring for higher prices, motivated by vanishing surpluses and higher prices of imports from the OECD nations. For instance, OPEC's press release, Number 9-77 of December 21, 1977, states: "The Conference considered the losses in purchasing power of export earnings of Member Countries resulting from continued imported world inflation. Furthermore, the Conference discussed the situation created by the weakening position of the U.S. Dollar and expressed also its deep concern, since oil prices are determined in that currency. The Secretariat was instructed to carry out a study on the subject and to propose remedial measures to safeguard

TABLE 6.5

Return on OPEC Commercial Investments, by Country, 1976-78

| Country | 1976 | 1977 | 1978 |
|---|---|---|---|
| Kuwait | 5.52 | 6.16 | 6.58 |
| United Arab Emirates | 4.17 | 4.69 | 5.00 |
| Saudi Arabia | 5.27 | 6.10 | 6.49 |
| Iran | 4.00 | 4.55 | 4.62 |
| Iraq | 6.40 | 7.14 | 10.00 |
| Libya | 3.00 | 3.00 | 3.33 |
| Venezuela | 2.44 | 2.20 | 2.00 |
| Average | 5.04 | 5.53 | 3.72 |

Source: *Economic & Political Weekly*, Annual 1980 (Bombay).

the interest of the Member Countries." The rationale for price increases was provided in OPEC's press release Number 6–78 of December 17, 1978, well before the major increases of 1979.

> The Conference reviewed the report of the Economic Commission Board and noted with great anxiety the high rate of inflation and dollar depreciation sustained over the last two years, and hence the substantial erosion in the oil revenues of the Member Countries and its adverse effects on their economic and social development. However, in order to assist the world economy to grow further, and also in order to support the current efforts towards strengthening the U.S. dollar and arresting the inflationary trends, the Conference has decided to correct only partially the price of oil by an amount of 10 percent over the year 1979. . . . The Conference, on the other hand, notes that, should inflation and currency instability continue, thus adversely affecting the oil revenues of the Member Countries and encouraging the wasteful use of this important, but depletable, resource, the Conference will find it imperative to adjust fully for the effects of such inflation and dollar depreciation.

Thus, the Iranian crisis merely provided a convenient opportunity to act in accordance with directions already set. The rapid elimination of current-account deficits in the OECD nations through the efficient use of petrodollar resources was the overwhelming reason for a real decline in OPEC revenues and a consequent desire for price increases. The term *exported inflation* has been used profusely in OPEC discussions and writings. For instance, Adnan Al-Janaby (1980) concludes that exported inflation, along with floating exchange rates, is the new means of international income redistribution. The author deflated OPEC imports for the period 1973–79 by the OPEC Import Price Index, and calculated that the $93.6 billion of OPEC imports from the industrialized countries in this period was worth only $25.2 billion in 1973 money. In many cases, according to Al-Janaby, inflation in services imported by OPEC, which account for about 30 percent of total OPEC imports, was even higher than that in merchandise.

Viewing these developments within the framework of global economic activity over the period 1973–79, we can conclude that the flow of capital played a major role in price movements for oil as well as for nonoil goods and services. Consequently, the second-round oil-induced adjustment, which began in 1979 and is yet to run its full course, may go the way of the first round, with some important differences. The effect of higher energy prices on overall consumer prices is relatively easy to establish, but their effect on aggregate output is difficult to measure. W. W. Hogan and A. S. Manne (1977) analyzed energy/gross-domestic-product interactions and established that the size of the energy sector in the U.S. economy is so small in relative terms that higher energy cost has a weak effect on output. Since energy inputs did not exceed 4 percent of gross national product in 1970, the authors contend that, if energy costs double and there is enough time for the economy to fully adjust to this change, the maximum effect on gross national product would be a 4 percent permanent loss. But, in fact, given time, the economy adjusts to the use of a different input mix, which depends on the extent of substitution achieved. The ultimate impact on output would then be determined by the speed and extent of this process. Consequently, the short-run effects of higher energy prices would be more severe than the long-run effects, if energy consumption is reduced immediately in accordance with inelastic demand curves representing consumer choices.

The empirical evidence of the two oil price shocks also clearly establishes the resilience of the economic systems in the OECD countries, which had lower rates of consumption over time without a substantial loss of output. For instance, in 1980 total output in these countries rose by a little over 1 percent, compared with 3.3 percent in 1979 and 3.9 percent in 1978. This is a better record than that of the first oil shock, after which output rose by only 0.6 percent in the first year and actually fell by 0.5 percent in the second year. Further, the effects of the second shock appear to have been synchronized in industrial countries, with recession setting in at the end of the first quarter of 1980. But the pace of each country was dissimilar in the period ensuing. By the beginning of 1980, the price of crude oil had increased by approximately 175 percent compared with the end of 1978. Oil imports into the OECD nations in 1980 were 10 percent less than in 1978, but its cost was equivalent to 2 percent of total national income. As before, the bulk of this payment was an accumulation of financial claims and not an actual transfer of resources in the shape of a rise in exports of goods and services. Unlike the policy responses after the first shock, this time around there was a serious attempt to limit the inflationary impact and align energy use with price. Hence recovery in growth has been slower than anticipated, but a greater effort at energy conservation and substitution of oil may leave the economies of these nations in better shape to counter the effects of future price increases. The impact of future oil price increases on gross national product in the industrial nations is, however, difficult to predict. Formal analysis based on quantitative estimation has to be combined with judgments on a host of global developments and variables. (See Jacoby and Paddock, 1981, for a description of future economic growth and oil prices.)

The impact of oil price increases on the developing countries can be measured in terms of their growth in exports (8.5 percent in 1979 and

1980) and imports (6 percent in 1979, 12 percent in 1980). Total export receipts in 1980 increased by 14 percent in unit values, but since import unit values rose by 21 percent, there was an actual deterioration in terms of trade by 7 percent, which largely offset the improvement achieved in the real trade balance. Balances on current account for the period 1973–81 are shown in table 6.6. The rising trend in deficits for the period 1978–81 brought about adverse impacts on their economies, and these appear to continue unabated, even though deficits in 1982–83 are likely to be lower. Nor is the outlook brighter given the prospect of economic recovery in the OECD nations and the possible accompanying rise in inflation. Indeed, the inflation in 1973–78 had unfavorable effects not only on the members of OPEC but on all developing countries. The aggregate effect would depend on whether increased earnings from exports to the OECD nations offset a decline in real export values resulting from worsening terms of trade. Largely as a result of these external changes, the economies of developing countries grew only 5 percent a year for the period 1979–81, compared with 6 percent a year for the decade ending 1978. A 5 percent growth rate may not seem catastrophic, but with a population growth of 2.5 percent a year, a 1 percent drop from expected per-capita increases has serious implications. Further, slowdown in growth has not been uniform; middle-income countries with access to bank financing have done relatively well, while the poorest have slowed down a great deal.

Although the past has not been as disastrous for developing countries as predicted, the future is of justifiable concern. Most developing countries started 1974 with a clean slate in foreign borrowings; this was full by 1978 and overfull by 1979–80. Oil importers, in particular, have such large current account deficits that they are unlikely to be able to pay their debts within the next ten years. The growth of external debts for the developing countries is shown in table 6.7. Of course, rising oil prices do not entirely account for all these debts; much of it is for investments in industrialization. In addition, exports from these countries declined with weak markets for manufactures, commodity prices declined, and world interest rates escalated, which increased the burden of payments on new debts, calling for further borrowing in many cases, merely to keep up repayment schedules.

Most governments lacked the strong domestic policies necessary to manage the situation, and their debts continued to increase. The banking system was reluctant to loan further amounts to ensure the sustenance of their political and economic systems. Over half of the $270 billion owed to private banks by developing countries in mid-1982 was the debt of Argentina, Brazil, and Mexico. Another 10 percent was owed by Chile and South Korea. Countries of east Europe owed

TABLE 6.6

Balances on Current Account, Developing Countries, Selected Years, 1973–81
(billions of dollars)

| Country Group and Item | 1973 | 1974 | 1975 | 1978 | 1979 | 1980 | 1981 | 1973–75 | 1978–81 |
|---|---|---|---|---|---|---|---|---|---|
| All developing countries[a] | | | | | | | | | |
| Oil trade balance[b] | −11.5 | −36.8 | −46.6 | −37.1 | −56.1 | −80.4 | −96.7 | −35.1 | −59.6 |
| Exports | −4.8 | −15.1 | −14.9 | −20.2 | −30.2 | −44.6 | −47.2 | −10.1 | −27.0 |
| Imports | 3.5 | 12.2 | 12.3 | 16.3 | 26.0 | 46.5 | 58.4 | 8.8 | 42.1 |
| Nonoil trade balance | −8.3 | −27.3 | −27.1 | −36.5 | −56.2 | −91.1 | −105.6 | −18.8 | −69.1 |
| Net services and private transfers | −5.7 | −17.3 | −25.5 | −11.2 | −14.9 | −16.8 | −25.0 | −19.8 | −13.8 |
| Gross investment income | −1.0 | −4.4 | −6.2 | −5.7 | −11.0 | −18.9 | −24.5 | −5.2 | −18.8 |
| Other services | −10.3 | −13.6 | −15.2 | −27.6 | −37.9 | −50.1 | −58.1 | −4.9 | −30.5 |
| | 9.3 | 9.2 | 9.0 | 21.9 | 26.9 | 31.2 | 33.6 | −0.3 | 11.7 |
| Oil-exporting developing countries | | | | | | | | | |
| Oil trade balance[b] | −2.6 | −5.1 | −10.0 | −7.6 | −8.0 | −10.6 | −14.8 | −7.4 | −7.2 |
| Exports | 0.4 | 1.9 | 2.9 | 5.8 | 10.7 | 22.2 | 30.7 | 2.5 | 24.9 |
| Imports | 1.3 | 5.2 | 5.7 | 8.7 | 15.0 | 28.0 | 37.3 | 4.4 | 28.6 |
| Nonoil trade balance | −0.9 | −3.3 | −2.8 | −2.9 | −4.3 | −5.8 | −7.0 | −1.9 | −4.1 |
| Net services and private transfers | −1.7 | −4.5 | −9.6 | −9.8 | −13.4 | −24.4 | −35.9 | −7.9 | −26.1 |
| Gross investment income | −1.3 | −2.5 | −3.3 | −3.6 | −5.3 | −8.3 | −9.6 | −2.0 | −6.0 |
| Other services | −2.3 | −3.3 | −3.5 | −6.8 | −9.1 | −11.6 | −13.2 | −0.8 | −6.4 |
| | 1.0 | 0.8 | 0.2 | 3.2 | 3.8 | 3.3 | 3.6 | −1.2 | 0.4 |
| Oil-importing developing countries[a] | | | | | | | | | |
| Oil trade balance[b] | −8.9 | −31.7 | −36.5 | −29.5 | −48.2 | −69.8 | −81.9 | −27.6 | −52.4 |
| Exports | −5.2 | −17.1 | −17.9 | −26.0 | −40.9 | −66.5 | −77.5 | −12.7 | −51.5 |
| Imports | 2.2 | 6.9 | 6.5 | 7.6 | 11.0 | 18.5 | 21.1 | 4.3 | 13.5 |
| Nonoil trade balance | −7.4 | −24.0 | −24.4 | −33.6 | −51.9 | −85.0 | −98.6 | −17.0 | −65.0 |
| Net services and private transfers | −4.0 | −12.7 | −15.7 | −1.4 | −1.6 | 7.3 | 10.4 | −11.7 | 11.8 |
| Gross investment income | 0.3 | −1.9 | −2.9 | −2.1 | −5.7 | −10.6 | −14.9 | −3.2 | −12.8 |
| Other services | −8.1 | −10.3 | −11.8 | −20.8 | −28.8 | −38.5 | −44.9 | −24.1 | |
| | 8.5 | 8.4 | 8.9 | 18.7 | 23.1 | 27.9 | 30.0 | 3.7 | 11.3 |
| Memorandum[c] | 3.22 | 10.49 | 11.05 | 12.84 | 19.02 | 30.90 | 35.20 | 7.83 | 22.36 |

[a] Excluding People's Republic of China.
[b] Based on volume of oil trade, according to U.S. Department of Energy data, valued at the oil-exporting countries' average oil export unit value.
[c] Average oil export unit value of oil-exporting developing countries, in U.S. dollars a barrel.

TABLE 6.7

Growth Rates of OECD and Developing Countries, 1973–82

|  | | Developing Countries | | | |
|---|---|---|---|---|---|
| Year | OECD Growth Rate (percent) | Growth Rate (percent) | Growth Rate, Western Hemisphere (percent) | Gross External Debt (millions of dollars) | Debt to Foreign Banks (millions of dollars) | Reserves[a] (millions of dollars) |
| 1973 | 6.1 | 6.7 | 8.4 | 110.0[b] | 35.0[b] | 25.9 |
| 1974 | 0.7 | 5.6 | 6.9 | 135.0[b] | 50.0[b] | 28.1 |
| 1975 | −0.2 | 4.2 | 3.2 | 165.0 | 62.7 | 27.1 |
| 1976 | 4.8 | 6.6 | 5.5 | 200.0 | 80.9 | 38.2 |
| 1977 | 3.8 | 5.4 | 5.0 | 250.0 | 94.3 | 52.1 |
| 1978 | 4.0 | 5.6 | 4.5 | 310.0 | 131.3 | 66.0 |
| 1979 | 3.1 | 5.0 | 6.7 | 365.0 | 171.0 | 76.7 |
| 1980 | 1.2 | 4.7 | 5.9 | 430.0 | 210.2 | 76.8 |
| 1981 | 1.5 | 2.5 | 0.5 | 505.0 | 253.5 | 74.7 |
| 1982 | −0.8[c] | 1.2[c] | −0.7[c] | 550.0[c] | 268.3[d] | 69.3[e] |

*Source*: Paul A. Volcker, "How Serious Is U.S. Bank Exposure?" *Challenge*, May–June 1983, pp. 11–19.

[a] Minus gold.
[b] Estimates done without benefit of Bank for International Settlements data, which are available beginning in 1975.
[c] Estimate.
[d] June.
[e] October.

TABLE 6.8

Bank Debts of Developing Countries, June 1982 (billions of dollars)

| Country | Total | U.S. Banks | Non-U.S. Banks | U.S. Share(%) |
|---|---|---|---|---|
| Mexico | 64.4 | 24.3 | 40.1 | 37.7 |
| Brazil | 55.3 | 20.7 | 34.6 | 37.4 |
| Venezuela | 27.2 | 11.1 | 16.1 | 40.8 |
| South Korea | 20.0 | 8.7 | 11.3 | 43.5 |
| Argentina | 25.3 | 8.6 | 16.7 | 34.0 |
| Chile | 11.8 | 6.3 | 5.5 | 53.4 |
| Spain | 23.7 | 5.7 | 18.0 | 24.1 |
| Philippines | 11.4 | 4.8 | 6.6 | 42.1 |
| Taiwan | 6.4 | 4.4 | 2.0 | 68.8 |
| Colombia | 5.5 | 2.7 | 2.8 | 49.1 |
| Greece | 9.7 | 2.7 | 7.0 | 27.8 |
| Yugoslavia | 10.0 | 2.5 | 7.5 | 25.0 |

*Source: Bank for International Settlements*, BIS semiannual maturity survey; U.S. Country Exposure lending Survey (data adjusted to a BIS basis) from Paul A. Volcker, "How Serious Is U.S. Bank Exposure?" *Challenge*, May–June 1983, pp. 11–19.

western banks a total of $64 billion. The vulnerability of U.S. banks in these financial transactions is indicated by their total claims of over $100 billion on the developing countries in mid-1982. Table 6.8 shows the claims of U.S. and non-U.S. banks on the major borrowing nations of the Third World in June 1982. Figures 6.2 and 6.3 show the changes that have taken place in the burden of debts on the developing countries between 1973 and 1981.

This crisis of unrepayable debt and consequent danger of bank failures has led to various suggestions on how the problem can be solved or at least postponed. Some radical leaders of Third World nations advocate a cancellation of a large part of these debts, on the grounds that the present situation has been brought about by the inequities of the international economic order and the world monetary system. More

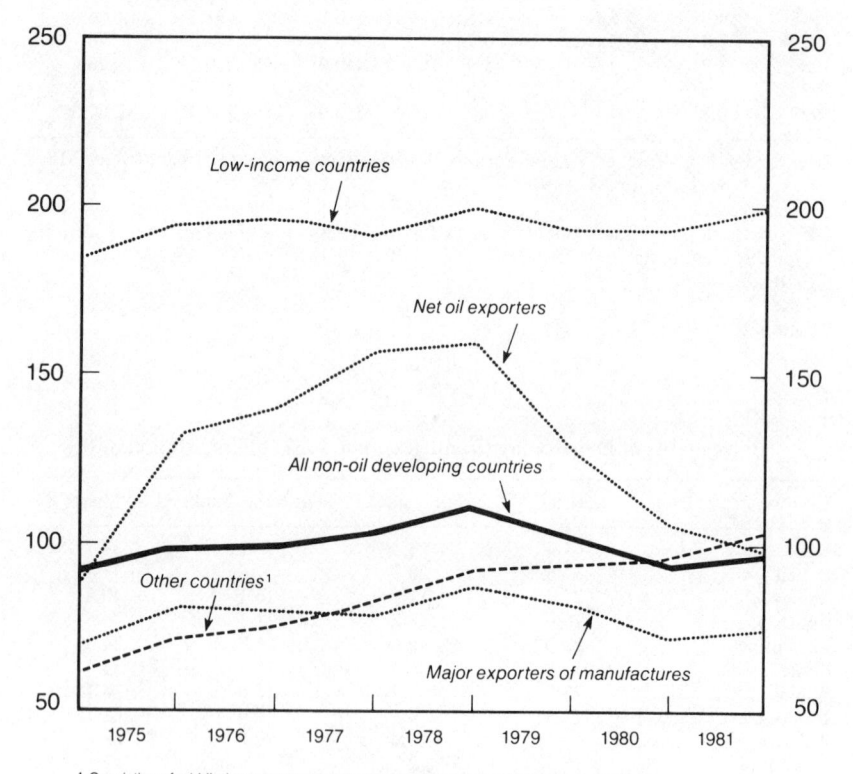

[1] Consisting of middle-income countries that, in general, export mainly primary products.

Figure 6.2 Oil-Importing Developing Countries, Ratio of Debt to Exports of Goods and Services, 1975–81

Source: International Monetary Fund, *World Economic Outlook* (Washington, D.C.: 1981).

responsible and pragmatic recommendations are to reschedule debts for those nations that cannot keep up repayments and to form policies and to restructure the economies of these countries. The International Monetary Fund has been pursuing readjustment programs in some of the worst affected nations, and agreement has been reached in Argentina, Brazil, and Mexico on economic policies directed at readjustment. The IMF's insistence on economic reforms in the Third World is often unpopular, but this organization has the authority and the reputation for political neutrality to prevail. These measures will, in the long run, nurse these fragile economies back to health, but the immediate future does not look bright. The burden of debts in the Third World has

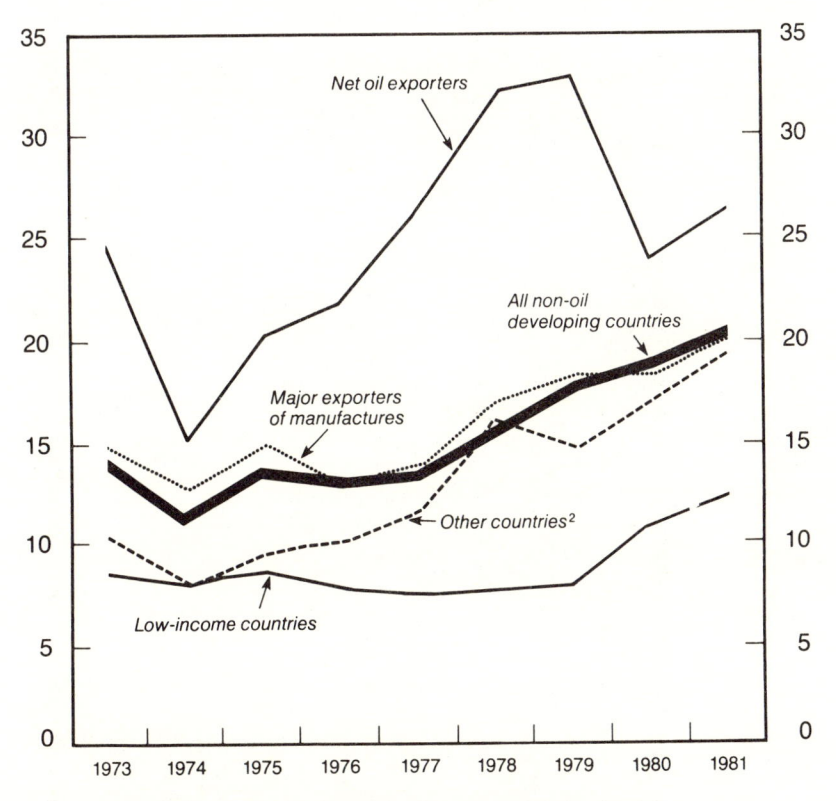

[1] Annual debt service payments as percentages of annual exports of goods and services.
[2] Consisting of middle-income countries that, in general, export mainly primary products.

Figure 6.3 Oil-Importing Developing Countries, External Debt Service Payments, 1973–81

*Source*: See figure 6.2

TABLE 6.9

Flows between Bank for International Settlement Reporting Banks[a] and Country Groups outside Reporting Area, 1974–80
(billions of dollars at current exchange rates)

| Country Group | Stocks at End-1973 | 1974 | 1975 | 1976 | 1977 | 1978 | 1979 | 1980 | Stocks at End-1980 |
|---|---|---|---|---|---|---|---|---|---|
| **OPEC countries[b]** | | | | | | | | | |
| Deposits | 16.0 | 26.5 | 7.5 | 12.5 | 12.5 | 3.5 | 37.0 | 41.5 | 159.7 |
| Borrowings | 6.5 | 2.5 | 5.0 | 9.5 | 11.0 | 17.5 | 7.5 | 6.0 | 70.0 |
| Net | 9.5 | 24.0 | 2.5 | 3.0 | 1.5 | −14.0 | 29.5 | 35.5 | 89.7 |
| Memorandum | | | | | | | | | |
| Foreign exchange reserves | 12.6 | 31.1 | 7.3 | 8.0 | 10.6 | −14.5 | 15.4 | 19.8 | 90.3 |
| Current account balances | — | 66.0 | 31.0 | 37.0 | 27.0 | −2.0 | 66.0 | 110.0 | — |
| **Other developing countries[c]** | | | | | | | | | |
| Deposits | 27.5 | 4.0 | 4.0 | 11.5 | 12.0 | 14.0 | 12.0 | 5.5 | 92.7 |
| Borrowings | 32.0 | 15.0 | 15.0 | 16.5 | 10.5 | 22.5 | 35.5 | 40.5 | 195.0 |
| Net | −4.5 | −11.0 | −11.0 | −5.0 | 1.5 | −8.5 | −23.5 | −35.0 | −102.3 |
| Memorandum | | | | | | | | | |
| Foreign exchange reserves | 21.2 | 0.6 | −1.3 | 10.6 | 10.0 | 11.9 | 7.8 | −1.9 | 58.9 |
| Current-account balances | — | −24.0 | −31.0 | −20.0 | −13.0 | −24.0 | −40.0 | −61.0 | — |
| **Developed countries[d]** | | | | | | | | | |
| Deposits | 27.0 | 0.5 | 5.5 | 1.5 | 4.5 | 8.5 | 7.5 | 5.5 | 50.1 |
| Borrowings | 23.0 | 7.5 | 10.0 | 12.5 | 12.5 | 5.5 | 7.5 | 15.0 | 85.6 |
| Net | 4.0 | −7.0 | −4.5 | −11.0 | −8.0 | 3.0 | — | −9.5 | −35.5 |
| Memorandum | | | | | | | | | |
| Foreign exchange reserves | 23.5 | −2.1 | −1.4 | 0.4 | 1.7 | 6.4 | 3.1 | 1.5 | 26.7 |
| Current-account balances | — | −17.0 | −19.0 | −21.0 | −22.0 | −7.0 | −6.0 | −14.0 | — |

|  |  |  |  |  |  |  |  |  |
|---|---|---|---|---|---|---|---|---|
| Eastern Europe |  |  |  |  |  |  |  |  |
| Deposits | 4.5 | 1.5 | 0.5 | 1.0 | — | 2.0 | 4.5 | 1.0 | 15.6 |
| Borrowings | 9.5 | 3.5 | 8.5 | 6.5 | 2.0 | 5.5 | 7.0 | 6.5 | 59.8 |
| Net | −5.0 | −2.0 | −8.0 | −5.5 | −2.0 | −3.5 | −2.5 | −5.5 | −44.2 |
| Memorandum |  |  |  |  |  |  |  |  |
| Trade balances[e] | — | −2.0 | −9.0 | −7.0 | −1.0 | −4.0 | 3.0 | 5.0 | — |
| Unallocated[f] |  |  |  |  |  |  |  |  |
| Deposits | 7.5 | 2.5 | 4.5 | 3.0 | 5.5 | 5.0 | 7.5 | 5.0 | 40.0 |
| Borrowings | 4.5 | 1.5 | 4.0 | 1.5 | 3.0 | 7.0 | 6.0 | 5.0 | 33.0 |
| Net | 3.0 | 1.0 | 0.5 | 1.5 | 2.5 | −2.0 | 1.5 | — | 7.0 |
| Total |  |  |  |  |  |  |  |  |
| Deposits | 82.5 | 35.0 | 22.0 | 29.5 | 34.5 | 33.0 | 68.5 | 58.5 | 358.1 |
| Borrowings | 75.5 | 30.0 | 42.5 | 46.5 | 39.0 | 58.0 | 63.5 | 73.0 | 443.4 |
| Net | 7.0 | 5.0 | −20.5 | −17.0 | −4.5 | −25.0 | 5.0 | −14.5 | −85.3 |

*Source*: Bank for International Settlements, *Annual Report, 1981*, p. 108.

[a] Up to 1977, the BIS reporting banks covered Belgium-Luxembourg, Canada, France, Germany, Italy, Japan, the Netherlands, Sweden, Switzerland, the United Kingdom, the United States, and the branches of U.S. banks in the Bahamas, Cayman Islands, Panama, Hong Kong, and Singapore. Thereafter they also covered Austria, Denmark, and Ireland, as well as certain trade-related items in domestic currency for banks in France and the United Kingdom not included before. The total flow figure for each year may not equal the difference between the amounts outstanding at the beginning and the end of the period as a result both of breaks in the series and of the method used for the calculation.
[b] Includes, in addition, Bahrain, Brunei, Oman and Trinidad and Tobago.
[c] Excludes offshore centers.
[d] Including up to 1977 Austria, Denmark, and Ireland, which are thereafter considered as part of the reporting area.
[e] Exports and imports f.o.b.
[f] Includes nonreporting offshore centers: (Barbados, Bermuda, other British West Indies, Lebanon, Liberia, Netherlands Antilles, and Vanuaatu (formerly New Hebrides), and international institutions other than the BIS.

reached unmanageable proportions and is bound to adversely affect growth in the future.

In a discussion of OPEC's global impact it is necessary to examine the main instrument for recycling petrodollars, namely, the international banking system. The Bank for International Settlements[3] reports that of the $212 billion increase in the international banks' external assets, approximately $140 billion represents crossborder claims within the BIS's reporting area. In actual terms, new lending within this area may be estimated at $80 billion in current dollars during 1980. Credits granted by the reporting banks to non-OPEC countries outside the BIS area rose by approximately $55 billion in 1980, or $62 billion if exchange-rate effects are excluded. New deposits received by the banks from developing countries were much lower than in 1979, and hence their net recourse was $35 billion, which was 50 percent higher than in 1979. Latin America alone borrowed $26.7 billion, leaving a relatively small amount for other regions.

Table 6.9 shows the activities of the reporting banks over the period 1973–80. Significantly, OPEC's place as the main supplier of funds has been taken by the OECD nations since 1975. In 1974, new deposits amounted to 40 percent of OPEC's current-account surplus, but in the following two years this share declined to 24 percent. In 1980, the share of OPEC surpluses deposited with the reporting banks stood at 38 percent, but by the beginning of 1981 this share was sharply reduced. However, funds continued to flow directly to the OECD countries from OPEC nations through nonbanking channels, and since a large share of these funds were invested in the banks by the OECD countries, in effect the banks still recycled large shares of OPEC surpluses to all oil-importing nations. In summary, the effect of international banking activities from 1974 through 1980 was that

1. The net increase in indebtedness of the oil-importing developing countries was $92.5 billion, almost half their cumulative current-account deficit of $210 billion during this period; their gross borrowings during the same period totalled $155 billion.
2. The pattern and extent of borrowing by these countries differed substantially from country to country, with the largest share going to Latin American countries and smaller amounts to east Asia, Southeast Asia, and other regions.
3. The developed countries outside the BIS reporting area borrowed a net sum of $37 billion and a gross amount of $70 billion, which represented a large proportion of their cumulative current-account deficits of around $105 billion.

3. *Fifty-first Annual Report* (Basle: 1981), pp. 104–11.

4. The east European nations borrowed $29 billion, almost twice their cumulative trade deficit during the period.
5. The banks in the BIS area loaned $158.5 billion, of which $82 billion came from the OPEC countries.

Demand for international credit from the commercial banking system is likely to remain high in the years to come, given the burden of oil imports and backlog of development plans in the developing countries. However, prospects are poor for increased lending by multilateral institutions like the World Bank and the International Monetary Fund and by the international commercial banks. First, some debtor countries have a poor history of repayment to commercial banks. Second, measures by banking regulatory authorities will limit growth of bank lending. Third, declining OPEC surpluses will reduce funds deposited in these banks. In the interests of commercial viability, banks may start attaching conditions to loans similar to those imposed by the IMF. The emphasis would be on loans to bring about structural changes in a country's economy.

Some other developments in recycling of petrodollars are generally predictable. It is likely that financial institutions in the Middle East which have gained valuable experience since 1974 will handle a larger share of future surpluses. They may diversify investments, perhaps, assuming larger risks in the effort to secure higher returns. Consequently, the recent trend in petrodollar flows, away from the United Kingdom and the United States, toward Germany, Switzerland, and Japan is likely to continue. Some of this capital may be recycled by banks in these countries to the developing countries, particularly if economic recovery in the OECD nations is slow in coming. At the same time, Arab banks may venture into direct lending to the Islamic nations of the developing world and to some middle-income countries like the Republic of Korea, Taiwan, and the stabler economies of Latin America. Some of the implications of these developments are discussed in subsequent chapters in the light of quantitative estimates of future surpluses and oil revenues.

In retrospect, perhaps the easiest means for providing capital to the developing countries for financing imports and development was for large-scale money creation at steady interest rates by the developed countries. If this inflationary path had not been adopted, then perhaps there would have been an enormous accretion of liquidity by the OPEC nations in the currencies of the West, since there was little recycling between OPEC and the developing countries. As it happens, the inflationary approach aided by large OPEC purchases of capital equipment, consumer goods, and services rapidly reduced the deficits of oil importers and the surpluses of exporters. Global inflation also facilitated the growth of developing countries' exports, because in an expansionary situation with low unemployment, the OECD nations were not opposed to higher imports from them.

But the lesson of these developments for the future is that those very policies that facilitated adjustment for oil importers imposed a burden on oil exporters, leading to the price increases of 1979–80. Even though inflation in the United States had been largely contained by early 1982, the prospects of inflationary pressures reemerging, once a strong economic recovery is under way, are present. Besides, inflation in most of Europe was higher than that in the United States, with no definite signs of a major improvement. An improvement in U.S. competitive strength and a reduction in inflation for a reasonable length of time would strengthen an already strong U.S. dollar. This would encourage a reduction in the rest of the world's dollar holdings, forcing a contraction in Eurodollar lending. The Deutsche mark and the yen would, therefore, have to be expanded to fill up the gap. But the strengthening of the dollar may slow down oil price increases and provide some welcome respite to oil importers. On the other hand, if U.S.inflation reemerges, then the dollar may weaken, and a repetition of the developments of 1977–78 may return, leading to possible ratcheting of oil prices. These prospects underline the importance of remunerative outlets for OPEC investments and the feedback effects between international economic developments and OPEC's pricing actions. Some scenarios predicting OPEC revenues and surpluses are examined in this framework in subsequent chapters.

CHAPTER 7

# THE INTERNATIONAL ECONOMIC ORDER

The oil price explosions of 1973–74 and subsequent years are the culmination of historic developments that are at the heart of the North-South dialogue and proposals for the establishment of a new international economic order. Essentially, the developing countries, historically exporters of primary products and commodities, have demanded higher prices for their exports and a reversal in declining terms of trade. Before the early seventies oil conformed closely to the trend of falling real prices for commodities traded internationally. Consequently, developing countries supported OPEC's pricing decisions as a matter of principle, even though the oil import burden has had a crippling effect on many of their economies. The inseparability of oil price increases and the North-South dialogue is in the interests of both OPEC countries and other developing countries. OPEC, beleaguered by opposition from the industrial nations, needed the support of other developing countries in pushing through oil price increases and in weathering the storm of opposition. Other developing countries find in the success of OPEC an inspiration for their own struggles to establish a new international economic order. Awed by OPEC's economic muscle, they have gone along with oil price increases publicly, even though they may have privately resented the effects on their economies.

The OPEC nations, in keeping with this solidarity, set up a special fund to help non-OPEC developing nations and support their demands for higher prices for raw materials. OPEC's press release Number 2-75 of January 26, 1975, states, "Convinced of the interdependence of nations and the need to promote solidarity among the peoples of the world through a genuine international cooperation, OPEC Member Countries welcome the dialogue between the industrialized countries and the developing countries and are, in this spirit, prepared to participate in an international conference, such as that proposed by the Government of France, which will deal with the problems of raw materials and development." OPEC has, accordingly, participated in major recent deliberations on this subject, including the 1981 summit at Cancun.

To ensure that the non-OPEC developing countries are not isolated, emotionally and politically, from them, OPEC countries have attempted to convince the world that the increase in oil prices was the least of the problems faced by these countries, and that their economic stagnation resulted from inequities between North and South. A publication by Paul Hallwood and Stuart Sinclair (1981) states that higher inflation, slower economic growth, and restrictions on imports into industrialized countries constituted external shocks for these developing countries far more serious than the first oil shock. These authors provide measures, calculated by Balassa, for assessing the impact of terms of trade losses and changes in export volume for the developing countries. These estimates are shown in table 7.1.

It can be seen from table 7.1 that direct losses from changes in terms of trade were quite substantial, and trade with the fuel-exporting nations was the largest component. Furthermore, these figures are aggregates for the period 1974–78 and do not convey the immediate impact of the oil price rise of 1973–74. It is evident from table 7.1 that the first oil price increase caused hardship for many developing countries, and some may voice opposition to price increases in the future. But the likelihood is that developing countries in Africa and Asia will continue to support OPEC, and even though nations of Latin

TABLE 7.1

Effect of Changes in Terms of Trade and Export Volume, Oil-Importing Developing Countries, 1974–78

| Country | Change in Exports | Effect of Terms of Trade with Industrial Countries | Effect of Terms of Trade with OPEC Countries |
|---|---|---|---|
| Argentina | − 33 | 270 | − 331 |
| Brazil | − 523 | 734 | − 3,107 |
| Colombia | − 136 | 215 | 16 |
| Mexico | − 257 | − 854 | 407 |
| Chile | − 76 | − 428 | − 252 |
| India | − 429 | 146 | − 1,200 |
| South Korea | − 451 | − 9 | − 1,309 |
| Singapore | − 284 | − 374 | − 350 |
| Taiwan | − 522 | 538 | − 901 |
| Yugoslavia | − 646 | − 1,259 | − 750 |
| Others | − 235 | − 471 | − 652 |
| Total | − 3,592 | − 1,492 | − 7,767 |

Source: Hallwood and Sinclair 1981, table 1.

America may be increasingly reluctant to do so, in balance, the developing countries will continue to support (and receive support from) OPEC and will present a solid front with OPEC in negotiations for a new economic order. This would ensure that OECD nations would join the dialogue, since they are vulnerable to oil availability and prices.

The grounds for the North-South dialogue were laid by the General Assembly of the United Nations in its Sixth Special Session in 1974, when it adopted the Charter of Economic Rights and Duties of States. Its major proposals include the following:

1. Adoption of integrated price supports for commodity exports of developing countries.
2. Indexation of prices of these commodity exports to prices of manufactured exports from industrial countries.
3. Formation of a development assistance fund of 0.7 percent of gross national product of industrial countries.
4. Linkage of the development assistance fund and special drawing rights by the International Monetary Fund.
5. Redeployment of some industries to developing countries.
6. Lowering import tariffs on manufactures from developing countries.
7. Development of an international food program.
8. Establishment of mechanisms for the transfer of technology to developing countries, separate from direct capital investment.

Two other major provisions of the charter deserve mention. One affirms each state's "full permanent sovereignty over its natural resources and economic activities," which supports, in spirit, a country's right to nationalize foreign property. The other provision upholds the right of commodity producers to associate in cartels, which other countries must refrain from trying to break up. The shadow of OPEC's example is very much present in this charter.

Of the main proposals listed above, the demand for an integrated commodity policy is by far the most significant, since it calls for organizational changes in the commodities' market for both exporting and importing nations. There are divergent views on the efficiency and desirability of such a policy. Harry G. Johnson (in Bhagwati 1977) is generally critical. According to him the objectives of such a policy are inconsistent: stabilizing prices in the short run, raising prices in every run, and achieving prices fair to both producers and consumers. In effect, it aims at parity of real income for producers of primary products and manufactured goods. The emphasis on integration, according to Johnson, is aimed at creating a monopoly, which producers would find appealing as a means for "organized exploitation of consumers." Yet the proposal aims for the support of the consumers. The thrust of this whole approach and the misplaced attention to commodity prices, in Johnson's view, represent a need to find an external scapegoat for the economic backwardness of the developing countries. Other economists, too, have commented unfavorably on the U.N. proposals (see, for example, I.M.D. Little in Amacher, Haberler, Willett 1979).

The United Nations Conference on Trade and Development (UNCTAD) has been pursuing the implementation of the proposals listed above. In 1975 it adopted an Integrated Programme for Commodities, with the following objectives related to commodity arrangements:[1]

1. Reduction of excessive fluctuations in commodity prices and supplies, taking account of the special importance of this objective in the cases of essential foodstuffs and natural products facing competition from stable-priced substitutes.
2. Establishment and maintenance of commodity prices at levels which, in real terms, are equitable to consumers and remunerative to producers.
3. Assurance of access to supplies of primary commodities for importing countries, with particular attention to essential foodstuffs and raw materials.
4. Assurance of access to markets, especially those of developed countries, for commodity-exporting countries.
5. Expansion of the processing of primary commodities in developing countries.
6. Improvement of the competitiveness of natural products vis-à-vis synthetics.
7. Improvement of the quantity and reliability of food aid to developing countries in need.

Priority is given to seventeen commodities (other than oil) of importance to export trade, particularly ten core commodities (cocoa, coffee, tea, sugar, hard fibers, jute and jute manufactures, cotton, rubber, copper, tin). Other proposals include (1) establishing a fund to finance international stocks of commodities, (2) setting up the stocks, (3) negotiating other measures necessary to attain the objectives of the program within the framework of international commodity agreements, and (4) improving compensatory financing to maintain stability in export earnings. The major components of the plan are the common fund and the international stockpile. Johnson describes the program "as

---

1. "An Integrated Programme for Commodities: Specific Proposals for Decision and Action by Governments," Report TD/B/C.1/193 (Geneva: Secretary-General of UNCTAD), October 28, 1975.

a demand for a massive investment of funds by the developed countries to underwrite experiments with and promotion of individual commodity-by-commodity agreements, experiments to the pursuit of which the UNCTAD Secretariat and its developing nation clientele are committed in spite of any uneasy half-recognition that international commodity agreements—aside from the difficulty of devising and operating them—are an extremely doubtful instrument for promoting economic development" (Bhagwati 1977, 243).

F.G.Adams and J.R.Behrman (1976) have investigated the impact of international buffer stocks on the commodities included in the UNCTAD program. They developed and estimated a series of supply-and-demand models for each commodity, using annual data for the period 1955–73. These models were then run to simulate the period 1963–75, driven by a set of assumptions the more important ones of which are summarized below:

1. The estimated equations for supply and demand remain reliable approximations of market behavior even when buffer stock operations are introduced.
2. The buffer stock managers operate with enough financial reserves to ensure average annual deflated commodity prices to be maintained within a specified band of ± 15 percent of the known secular trends for 1950–75.
3. The potential gains from pooling finances across commodities are measured by the extent to which total financial requirements are reduced by using funds from the sales of some commodities to purchase others.

The authors established that large gains do accrue from price stabilization of the UNCTAD type; one set of assumptions show a possible producer gain of $21 billion. Further benefits include revenue and income stabilization for the producing countries, and there are large savings from financial pooling. Further work by Behrman (1978) adds greater support to his initial qualified endorsement of the UNCTAD proposals.

A common fund under the Integrated Programme for Commodities (IPC) has finally been established after prolonged discussions, with contributions from individual countries and OPEC. But it has hardly started functioning effectively, and controversy on the efficiency of the arrangement continues unabated. On the basis of research done on the subject, the major issues related to the IPC can be summed up as follows:

1. The IPC proposal and the operation of a buffer stock for commodities can prove useful in achieving price and revenue

stabilization to the advantage of producers and, to a lesser extent, to the advantage of consumers as well.
2. The benefits and costs of the scheme vary considerably across commodities, and the administrative and other indirect costs cannot justify the application of the scheme equally to all commodities identified by UNCTAD. A more efficient and pragmatic approach would be to initially apply the buffer stock operations to those commodities that exhibit the greatest fluctuations.
3. There is little evidence of net benefits from an integrated program, and the possibility of inequities between producers of different commodities is very strong.
4. The greatest danger of a large-scale program of this variety, however, lies in the long-run institutionalization of an arrangement that could experience increasing costs and declining benefits as individual countries stabilize prices by establishing their own buffer stocks and minimize losses by diversifying production and exports.
5. It is naive to believe that UNCTAD's program can be used as an instrument for arresting the decline of terms of trade for major commodity producers. As long as consumer nations are at the losing end of net transfers through higher commodity prices, the scheme would remain unacceptable and, therefore, politically infeasible. In the long run, there is no alternative to rising prices being signals for scarcity, which would come about only through a reallocation of resources to other forms of production.

There is, in conclusion, considerable merit in finding alternative mechanisms for effecting resource transfers other than by providing artificial props for commodity prices (as opposed to price stabilization, which has great merit), since any such prop would rest on agreements between parties with conflicting interests. These alternative mechanisms should go beyond the limited role of price stabilization (for which buffer stocks are suitable), transferring resources from the richest to the poorest countries in the most efficient manner and involving all parties.

Much has been written, in this spirit, on the mutuality of interests between North and South. A recent plea for a change in priorities is that of the Brandt Commission.[2] On the subject of commodity prices, for instance, the commission makes the point effectively that price stabilization is in the interests of both North and South, since it would have a favorable effect on investments and long-term supplies and in

2. Independent Commission on International Development Issues, *North-South: A Programme for Survival* (London: Pan Books, 1980).

holding down inflation in countries of the North.[3] Similarly, the benefits of increased development aid from North to South are shown as accruing in the form of increased demand for exports from the developed countries and in higher marginal returns from capital invested, since these returns are likely to be higher on the margin in the capital-scarce developing countries than in the developed nations. The report of the commission also makes frequent references to the benefits for the North from stability in international relations. The arguments made by the Brandt Commission are well presented, but it has not drawn much attention. Through unfortunate timing, the report came out in the wake of the second oil shock of 1979–80, which was soon followed by recession in the North. The Reagan administration, which came into office in 1981, had no sympathy with development assistance that did not promote the strategic interests of the United States.

On balance, the North-South dialogue has not been going well, and both sides are responsible. The North continues its insular approach of "business as usual," opposing any structural change in the international market for goods and services. The South politicizes its demands (like a trade union, which is essentially what it is), trying to achieve instantaneously, through drama and rhetoric, what it should achieve through careful homework and persistent negotiations. There are also changes and reforms the South could have brought about in its own hemisphere, which would have made better use of the existing capital flows from the North, however meager. According to Mahboob ul Haq, most developing countries regard the North-South debate as a soft option, ensuring concessions and benefits for them without any internal effort.[4] But the record of the developing countries in seizing the initiative in their own spheres of action is mixed, and global inaction in restructuring the international economic order does not spell doom and misery for all of them but for only the poorest.

During the past decade, major changes have taken place in the structure of merchandise trade, the directions of which are brought out in table 7.2. It can be seen that the performance of the middle-income oil-importing nations in increasing the share of manufactures in their total exports has been noteworthy. This shift has been made possible by a relatively large increase in the aggregate volume of trade in manufactures for the world as a whole. The composition and growth of world merchandise trade is shown in table 7.3. Manufactured exports, as can be seen in this table, grew faster in the seventies than did primary

---

3. Typically, a sharp rise in the price of a commodity has inflationary effects on the price of goods in the industrial nations, causing a price spiral, based on price expectations, even beyond the time the price of the commodity has declined.

4. Mahboob ul Haq, "Negotiating the Future," *Foreign Affairs* 59(2), 398–417.

TABLE 7.2

Merchandise Trade, Oil-Importing Countries, 1970 and 1980 (percent)

| Country Group | Export-GDP Ratio | Exports Manu-factures | Exports Nonfuel primary | Imports Manu-factures | Imports Food | Imports Fuel |
|---|---|---|---|---|---|---|
| | | | 1970 | | | |
| African oil importers | 23 | 11 | 86 | 77 | 11 | 9 |
| Asian oil importers | 7 | 54 | 43 | 64 | 21 | 5 |
| Middle-income oil importers | 22 | 33 | 58 | 69 | 12 | 10 |
| | | | 1980 | | | |
| African oil importers | 16 | 9 | 80 | 51 | 16 | 31 |
| Asian oil importers | 9 | 47 | 50 | 38 | 14 | 39 |
| Middle-income oil importers | 24 | 46 | 36 | 53 | 11 | 28 |

Source: World Bank 1981b, table 3:4.

TABLE 7.3

Composition and Growth of World Merchandise Trade, 1970–80

| Item | Total Merchandise | Fuels | Nonfuel Primary Products | Manufactures | Gold |
|---|---|---|---|---|---|
| Value, 1980, billions of current dollars | 2,133 | 535 | 400 | 1,170 | 27 |
| Value, 1980, percent of total | 100 | 25 | 19 | 55 | 1 |
| Increase in value, 1970–80, billions of current dollars | 1,818 | 507 | 312 | 973 | 26 |
| Increase in value, 1970–80, percent of total | 100 | 28 | 17 | 54 | 1 |
| Higher prices as a percent of increased value | 87 | 98 | 82 | 81 | 101 |
| Increase in volume, 1970–80, percent | 74 | 29 | 64 | 96 | −4 |

Source: World Bank 1981b, table 3:1.

exports, and the developing countries expanded their exports of manufactured goods faster than in the sixties, despite slower growth in the industrial nations, which are the main markets for these exports. Changes in merchandise trade in the seventies are shown in table 7.4. Middle-income oil exporters increased their share of manufactures exports and reduced their share of manufactures imports, in large part

TABLE 7.4

Share of Merchandise Exports, by Country Group, 1960 and 1980 (percent)

| Country or Group | Fuels, Minerals, and Metals 1960 | Fuels, Minerals, and Metals 1980 | Other Primary Commodities 1960 | Other Primary Commodities 1980 | Textiles and Clothing 1960 | Textiles and Clothing 1980 | Machinery and Transport Equipment 1960 | Machinery and Transport Equipment 1980 | Other Manufactures 1960 | Other Manufactures 1980 |
|---|---|---|---|---|---|---|---|---|---|---|
| Low-income countries (excluding China and India) | 8 | 9 | 83 | 62 | 4 | 21 | — | 2 | 5 | 6 |
| China and India | — | 20 | — | 30 | — | 18 | — | 5 | — | 27 |
| Middle-income oil exporters | 40 | 78 | 48 | 15 | 1 | 2 | — | 2 | 3 | 3 |
| Middle-income oil importers | 15 | 12 | 68 | 34 | 5 | 13 | 2 | 14 | 10 | 27 |
| Industrial countries | 11 | 13 | 23 | 15 | 7 | 5 | 29 | 35 | 30 | 32 |
| Capital-surplus oil exporters | — | 98 | — | — | 0 | — | 0 | 1 | 0 | 1 |
| Nonmarket industrial countries | 18 | — | 33 | — | 3 | — | 34 | — | 21 | — |

*Source:* World Bank 1983, table 10.

because of the increase in share of fuels in total imports. This development reversed a trend evident in earlier decades, with exports from these countries growing more slowly than those of the industrial nations, largely because the income elasticity of demand for primary products is low. The faster rate of growth has come about through changes in the structure of production and is essentially supply driven. But the performances of these countries have been very uneven, the fastest growth and diversification having come about in east Asia—the Republic of Korea, Taiwan, Singapore, and Hongkong.

While manufacturing has been a major source of growth in exports from the developing countries, the increase in oil prices brought about a radical change in export earnings in the low-income group, particularly Indonesia (and Nigeria, which is now classified as a middle-income country). Developing countries' share of agricultural exports, on the other hand, has declined over time, resulting in some cases from a concentration on tropical produce such as coffee, cocoa, tea, and bananas, and in other countries from a shift from agriculture to manufacture for domestic markets. The demand for the commodities mentioned has not kept pace with growth in income of the developed nations, and hence coffee and banana growers in Latin America, cocoa growers in Africa, and tea growers in south Asia have not benefited from the boom in international trade that has favored other exports. Consequently, growth of agricultural exports from developing countries in recent years has been about half that of the developed countries, which have concentrated on food grains, for which demand has grown rapidly.

In the case of nonfuel minerals, developing countries approached growth rates close to 5 percent a year in the last two decades, but fluctuations in demand continue to be substantial. As it happens, a number of these minerals are exported by the least developed countries (for example, sub-Saharan Africa), which are totally dependent on exports of these commodities for their income. Consequently, fluctuations in demand cause large fluctuations in their income. Attempts at diversification have not yet borne fruit. The unfavorable impact of price and demand fluctuations on exporting nations is likely to continue, unless the UNCTAD program for price stabilization takes effect for all such commodities.

The major potential for growth in exports for the developing countries is manufactures. The 1979 pattern of trade in manufactures for developing countries is given broadly in table 7.5, which clearly brings out the importance of the industrial countries in their exports of manufactures. By comparison, trade in manufactures among developing countries is small, but exports from developing countries to other

TABLE 7.5

Composition of Trade in Manufactures, Developing Countries, 1979

| Manufactures | Exports to and Imports from Developing Countries — Billions of Dollars | Percent | Developing Countries — Exports to Industrial Countries — Billions of Dollars | Percent | Imports from Industrial Countries — Billions of Dollars | Percent | Imports from Developing Countries, Share of Total (percent) |
|---|---|---|---|---|---|---|---|
| Chemicals | 3.640 | 12 | 2.921 | 5 | 26.125 | 14 | 12 |
| Machinery and transport equipment | 9.414 | 29 | 11.562 | 20 | 103.982 | 54 | 8 |
| Other manufactured goods | 18.555 | 59 | 42.596 | 75 | 61.654 | 32 | 22 |
| Total or average | 31.609 | 100 | 57.079 | 100 | 191.761 | 100 | 14 |

Source: United Nations, Yearbook of International Trade Statistics 1979/80.

developing countries and to the centrally planned economies have been growing at over 10 percent a year, keeping pace with growth for manufactured exports as a whole. Growth rates by individual products, too, have generally been in excess of 10 percent a year.

In a recent World Bank publication, Hollis B. Chenery and Donald B (1979) Keesing (1979) analyze the changing composition of exports from the developing countries and establish that the growth of their manufactured exports has been a success story, for which credit should go to the exporting countries concerned, even though liberalized trade policies of the industrial countries in the sixties and seventies were an important influence. But the outlook for the future is clouded. First, there is the danger of a few successful developing countries taking advantage of new opportunities at the cost of developing countries new at the game. The authors also feel that where quotas are likely to continue, as in the case of textiles and some agricultural products, it would be desirable to modify them to favor the least developed and poorest nations. But there is in such an approach the quandary of efficiency versus equity; and also, competition among developing countries might promote policies that are more outward looking. In reality, developing countries will perform well only if the markets for their products grow. Unfortunately, the eighties have started with worldwide recession followed by protectionist measures by the developed countries, which run counter to the philosophy of free trade. According to Bela Balassa, current protectionism, generally nontariff restrictions on trade and government aid to domestic industries, is inferior to the old form of protection, which was essentially tariffs. Tariffs, as Balassa rightly comments, are after all instruments of the market economy and are therefore more efficient economically than nonmarket measures. Protectionist measures taken by industrial nations, over and above being harmful to developing countries, lead in the long run to income losses to the developed countries themselves, since resources will not be allocated in accordance with the principle of comparative advantage. But economic doctrine and logic propounded in international forums often give way to political expediency. Trade restrictions are, therefore, likely to continue, and indications of change are not apparent. The Tokyo Round of tariff concessions, while opening on a hopeful note for the developing countries, took on protectionist overtones from the North once the negotiations got going, apparently the result of the adverse economic trends of the seventies. In the meantime, in the United States steelworkers and their employers protested the "dumping" of foreign steel, and a financially sick Chrysler Corporation, unable to stand competition from imports, was bailed out by the U.S. government.

Thus the outlook for a major growth in the 1980s in exports from the developing countries to the developed countries does not appear bright. Exporters of commodities are likely to face even more unfavorable prospects, with sluggish growth in demand and large price fluctuations. Even though the evidence regarding the operation of buffer stocks and the establishment of a common fund is mixed,[5] major benefits are likely from such a price stabilization mechanism, at least for those nations whose commodity export earnings form a large part of their national incomes. The flexibility and specificity of the buffer stock approach give it great merit over the financing scheme of the International Monetary Fund, with its rigid conditions and constraints.

Given the realities of narrow self-interest and the lack of progress in the North-South dialogue (as highlighted in the sterile Cancun summit), a reappraisal and redefinition of objectives and a revised approach to the problem may be long overdue. The dialogue may, therefore, be resumed by accepting and emphasizing the following facts:

1. Global interdependencies in the economic field will get stronger in the future, and their recognition is of benefit to all nations.
2. The North must follow an enlightened approach, sensitive to the genuine needs of the South; had such an approach been taken in the pricing of oil before 1973, the world might have been spared the shocks that followed.
3. The South could substantially reform their own systems and orient their economies toward greater and freer participation in international trade.
4. Both North and South should expand the common fund to underwrite buffer stocks for more commodities. Financing of the fund should come in large measure from the richer nations of the South, namely OPEC members. The price stabilization program should not be complicated by using it as an instrument for price increases of commodities. In the long run, any artificial propping of prices would be inefficient and would harm the development and diversification of export-based economies in the developing countries.
5. The South should seek greater liberalization of trade with the North as a means for mutually beneficial growth in both regions.
6. Resource transfers to the poorest countries in the world should be stepped up, using bilateral and multilateral channels, such as the International Monetary Fund and the World Bank. OPEC's share in such efforts should be much larger and more evenly spread.

5. For an excellent summary of the research on this subject and the issues involved, see Gordon Smith, "Commodity Instability: New Order or Old Hat?" in Amacher, Haberler, and Willett 1979.

In general, the North-South dialogue needs to be reinitiated with different approaches by both sides. And OPEC, which has thus far been a passive participant, should play a more active role as a member of the South and as the supplier of a commodity of critical importance to the North. Unfortunately, OPEC's sudden rise from rags to riches has not given it time to mature into a role commensurate with its economic power. Also the muddled and violent politics of the Middle East has diverted its energies and strengths from a positive global program. Furthermore, the expected decline in OPEC external aid as a consequence of lower surpluses in the next few years may not allow any change in geographical emphasis. But perhaps a decline in oil revenues will reduce OPEC's religious bias and lead to more even-handed aid to other developing countries.

## CHAPTER 8
# AGENDA FOR CHANGE

The most compelling reason for attention to the North-South problem is the unprecedented financial difficulties faced by the developing countries. Even though, in the seventies, their growth rates were higher than expected, many have gone beyond their capacities to borrow from the international system. The problem has been compounded, of course, by the oil price increases of 1979–80.

### *The Brandt Commission Report*

Coinciding with the timing of the 1979–80 oil price increases, the Brandt Commission published its report on the problem, putting forward a set of well-reasoned (and even more well-intended) recommendations for the consideration of world leaders.[1] The moving spirit behind the commission's recommendations is the globalization of policies. The report argues that there is a great mutuality of interests between North and South, and if the actions proposed are not taken, then catastrophe will engulf the poor and decay will take root among the rich. The major challenge in pushing through reforms is to show that they are of material and economic value to those who have to make them. The opening sentence of the report states that the commission could "contribute to the development of worldwide moral values." But the concrete recommendations of the commission have to go beyond the refashioning of moral values and deal with issues on which some action could be initiated with relative ease, for redressal of the serious imbalances facing the world. Essentially, the commission argues in favor of a kind of global welfare state, which provides relief to the more vulnerable members of the international community through a fund of resources generated by taxing international trade and arms exports, exploiting the oceans' resources, and so on. The actual establishment

---

1. Independent Commission on International Development Issues, *North-South: A Programme for Survival* (Cambridge, Mass.: MIT Press, 1980).

and operation of such a global welfare state is of course contingent on agreement from all nations of the world and the resolution of various administrative and organizational problems. But there are a number of specific issues not directly linked with this central theme which can be discussed here.

The commission argues, for instance, in favor of international monetary reform, which is important in view of the recent fluctuations in exchange rates. The commission recommends a much larger role for special drawing rights (SDRs), which it views as the only stable and permanent currency. The International Monetary Fund is, therefore, urged to increase its allocations of this medium to the developing countries. The commission also feels that special drawing rights could be distributed more equitably, on the basis of criteria different from the existing quotas of IMF member countries. The Third World nations have for some time been asking for a revised set of criteria in establishing quotas for representation on the board of the IMF, and this demand has been put forward in the various discussions of the Group of Seventy-seven and the Group of Twenty-four (groups of developing countries set up for negotiating changes in the international economic order). The commission further recommends that countries with surpluses be encouraged to make long-term loans to developing countries that suffer deficits, and that the IMF reduce its allocations to nations with surpluses to divert a larger share to developing countries with deficits.

The commission also states that by making the SDR the principal reserve asset, holders of international reserve currencies, especially the dollar, will have greater freedom to diversify their portfolios. Reference has been made in this regard to the IMF's efforts for a "substitution account," which would enable holders of unwanted dollars to change them for SDRs or liquid SDR-denominated assets. Foreign exchange would be deposited in this account, to be administered by the IMF in return for SDR-denominated claims. The commission also suggests that the IMF use its gold holdings (which amount to 100 million ounces) either for (1) further sales, in which case the profits would subsidize the interest on loans to low-income developing countries, or (2) as collateral to borrow against in order to lend to developing countries, particularly those in the middle-income category.

While suggesting the expanded use of the SDR, the commission notes that the valuation will have to be on a basis that will assure predictability and stability, and that present limits on its use should be relaxed, making it as usable as other forms of central bank reserves. Such changes would encourage the valuation of items of international trade in terms of the SDR. This suggestion is interesting from the point of view of commodity prices. One of OPEC's concerns has been fluctuations in the value of the dollar, particularly the decline of the late seventies. If the SDR were an international currency, price changes to compensate for fluctuations in revenues brought about by exchange rate movements would not be necessary to the extent they are now.

Other recommendations of the commission relate to development finance, concessional finance, and other forms of assistance to the developing countries. It also suggests that a new world development fund be set up, which would complement the lending activities of existing agencies. It points out that while in 1960 about 60 percent of financial flows to the developing countries were in the form of official development assistance, by 1977 more than two-thirds was provided at commercial rates in the form of private bank loans, direct investments, or export credits. As terms get increasingly stringent, the Third World will have increasing difficulty meeting payments. The commission advocates an increased flow of development finance for food production, exploration for and exploitation of energy and mineral resources, and domestic processing of commodities. These "massive transfers" are described as in the interests of the countries of the North as well, because they would expand global markets for manufactured goods and revitalize the world economy.

The commission puts forward the concept of universal responsibility and automatic revenues for development assistance, which would have nothing to do with political considerations. These revenues would come from (1) a tax on all but the least developed countries and based on a sliding scale related to national income; (2) levies on international trade, arms exports and international travel; and (3) profits from new global enterprises, such as the new institution for exploiting ocean resources. A world development fund, having universal membership, would fill gaps in development assistance and provide an institutionalized conduit for the automatic revenues generated from the three sources. The money would be channelled through regional and subregional institutions, and an attempt would be made to cofinance its programs, as appropriate, through the World Bank and regional development banks.

As would be expected, the report of the commission has been criticized by various individuals and organizations. For instance, an article in the *Wall Street Journal* found the whole effort of the commission misdirected (Krauss 1980). It contends that, "Written by Social Democrats, the report closely follows the party line throughout its 304 pages." The emphasis on income transfer rather than income creation is seen to be in line with the Social Democrats' "unshakable faith in the curative powers of income transfer." The author of this newspaper piece also appears skeptical of the commission's belief that military spending is an obstacle to economic development. Income transfers, the *Wall Street Journal* says, are, in effect, a zero sum game—for one to gain from it another must lose. Government coercion, therefore, is needed, and since there is no world government, "nation states cannot be coerced into the international income transfer the Social Democrats desire." The commentary further criticizes the implied incentive to inefficiency that the recommendations would provide, such as loans on lenient terms. "There is a reason a country needs the loan most, and that reason often is irresponsible economic management. By rewarding economic irresponsibility, the Brandt report encourages it." The article ends by stating, "If the Social Democrats who wrote the Brandt report took time off from lecturing others on how to manage world affairs to put their own house in order, the Third World—among others—would benefit immensely."

I quote extensively from this newspaper review only because it epitomizes the philosophical opposition to the spirit and contents of the report evident in influential circles in many countries. Undoubtedly, there are basic flaws in the reasoning in many parts of the report, and, in general, it has not addressed the weaknesses of existing institutions and structures. Besides, it pays little attention to the practicalities of selling these programs and their implied sacrifices to ordinary citizens in industrial countries. But, perhaps, an emphasis on philosophy was both deliberate and necessary. If it had been preoccupied with implementation and operation, the reader's attention might have remained confined to these relatively unimportant matters without consideration of major policy issues, which is the main purpose of the commission. Besides, a report of this nature cannot possibly answer all questions and solve all problems connected with the implementation of its findings; it only provides a basis for dialogue on the restructuring of the global economic order. For the same reason, one cannot expect uniform responses and reactions to the contents of the report. Those who have a deep faith in the efficiency and fairness of the present system view the agenda for reform as misplaced, woolly-headed, and unworkable. In their view a global welfare state is as much of a distortion as a national welfare state.

The importance of the Brandt Commission and its recommendations is that it comes at a time when the international economic order is at the crossroads. The historic relationship of commodity exporters and the developed nations—consumers of commodities—has changed radically, with one group of commodity exporters, that is, the oil-exporting nations, having found sudden wealth and economic power. The increase in oil prices has left the other commodity exporters with severely limited options and heavier financial burdens. However, the objections of those who favor an ethic of self-help and advocate using private sources of financing deserve discussion.

## *Financing Development in the Third World*

The history of financing the developing nations by private banks in the West goes back to the end of the 1960s. Before that, most capital flows to these countries were in the form of grants and loans from official bodies, generally on concessional terms. In fact, in 1969 about 55 percent of the outstanding debt of the developing countries was from official sources, and the balance consisted mainly of officially guaranteed suppliers' credits. About the same time, U.S. and European banks began to see the benefits of overseas lending, and they extended loans to developing countries. After 1973, the magnitude of current-account surpluses and deficits went up sharply. A marked change took place, consequently, in the pattern of financing for the Third World, particularly for the oil-importing countries, whose demands for external finance went up sharply with the increase in oil prices. Current-account deficits and sources of finance for these countries are shown in table 8.1. The share of commercial loans went up sharply between 1970 and 1980 for the middle-income countries in particular. Commercial loans to the low-income countries, on the other hand, remained relatively small, implying unfulfilled demand for external finance during this period. This is evident from the fact that, in 1980, the current-account deficits as percentage of gross national product of the two groups were 4.5 and 5.0, respectively, indicating in percentage terms a similar level of need for external finance to cover deficits. In 1980, low-income countries received capital flows to cover 72.5 percent of their deficit, as opposed to 80.8 percent for the middle-income countries. Official development assistance for low-income countries grew rapidly in the period 1973–75 but slowed down considerably since.

Borrowing in the syndicated loan market by different groups of countries is shown in table 8.2. In general, most developing countries found borrowing from the private market a desirable option, mainly because credits are arranged quickly, and there is complete flexibility in the use of these funds. Besides, private banks have, in recent years, been the only lending organizations with adequate funds available to meet the immediate needs of the developing countries. Conditions for syndicated loans can be measured by spreads over the London interbank offered rate (Libor) and the average maturities of credit granted. These two measures for the period 1975–81 are shown in table 8.3. It can be seen that spreads were high and maturities low in the

TABLE 8.1

Current-Account Deficit and Finance Sources, Oil-Importing Developing Countries, Selected Years, 1970–80
(billions of 1978 dollars)

| Item | Low-Income Countries |  |  |  | Middle-Income Countries |  |  |  |  |
|---|---|---|---|---|---|---|---|---|---|
|  | 1970 | 1973 | 1975 | 1978 | 1980 | 1970 | 1973 | 1975 | 1978 | 1980 |
| Current-account deficit[a] | 3.6 | 4.9 | 7.0 | 5.1 | 9.1 | 14.9 | 6.7 | 42.8 | 20.4 | 48.9 |
| (As percent of GNP) | (1.9) | (2.4) | (3.9) | (2.6) | (4.5) | (2.6) | (1.0) | (5.5) | (2.3) | (5.0) |
| Net capital flows | 3.7 | 3.8 | 7.1 | 5.1 | 9.0 | 14.8 | 12.4 | 42.8 | 20.4 | 49.0 |
| Official development assistance | 3.4 | 4.1 | 6.6 | 5.1 | 5.7 | 3.3 | 5.3 | 5.3 | 6.5 | 7.9 |
| Private direct investment | 0.3 | 0.2 | 0.4 | 0.2 | 0.2 | 3.4 | 5.1 | 3.8 | 4.6 | 4.5 |
| Commercial loans | 0.5 | 0.6 | 0.8 | 0.9 | 0.7 | 8.9 | 13.7 | 21.0 | 29.4 | 27.1 |
| Changes in reserves and short-term borrowings[b] | −0.5 | −1.1 | −0.7 | −1.1 | 2.4 | −0.8 | −11.7 | 12.7 | −20.1 | 9.5 |

*Source:* World Bank, Economic Analysis and Projections Department.
[a] Excludes net official transfers (grants), which are included in capital flows.
[b] A minus sign indicates an increase in reserves.

TABLE 8.2

Borrowing in the Medium-Term Eurocurrency Credit Market, 1973–81 (billions of dollars)

| Country Group | 1973 | 1974 | 1975 | 1976 | 1977 | 1978 | 1979 | 1980 1st half | 1980 2nd half | 1981 1st half[a] |
|---|---|---|---|---|---|---|---|---|---|---|
| All borrowing countries | 17.2 | 28.5 | 20.6 | 28.7 | 34.2 | 73.7 | 70.2 | 29.0 | 41.4 | 32.8 |
| Developing countries, | 8.6 | 10.4 | 13.0 | 18.1 | 20.1 | 38.2 | 43.2 | 16.7 | 20.5 | 19.2 |
| percent of total | (41) | (37) | (63) | (63) | (59) | (52) | (62) | (58) | (50) | (59) |
| Oil-importing developing countries | 4.4 | 7.7 | 7.1 | 11.8 | 10.4 | 19.6 | 26.5 | 11.7 | 13.4 | 14.0 |
| percent of total | (21) | (27) | (35) | (41) | (31) | (27) | (38) | (40) | (32) | (43) |
| OEDC countries, | 4.2 | 2.7 | 5.9 | 6.3 | 9.7 | 18.6 | 17.2 | 5.0 | 7.1 | 5.2 |
| percent of total | (20) | (9) | (29) | (22) | (28) | (25) | (25) | (17) | (17) | (16) |

*Source*: World Bank, Debtor Reporting System.

[a] First half of year.

TABLE 8.3

Terms for Syndicated Eurocredits, 1975–81

|  |  |  |  |  |  | 1980 |  | 1981[a] |
|---|---|---|---|---|---|---|---|---|
| Country Group | 1975 | 1976 | 1977 | 1978 | 1979 | 1st half | 2nd half | 1st half |
|  | Spread over Libor[b] (percent) |||||||| 
| Oil-importing developing countries | 1.73 | 1.76 | 1.85 | 1.25 | 0.77 | 0.85 | 0.95 | 1.09 |
| Industrialized countries | 1.47 | 1.34 | 0.97 | 0.95 | 0.54 | 0.57 | 0.53 | 0.50 |
|  | Maturity (years) ||||||||
| Oil-importing developing countries | 5.5 | 5.5 | 6.2 | 9.6 | 9.7 | 9.2 | 7.8 | 8.0 |
| Industrialized countries | 5.9 | 6.2 | 6.7 | 7.5 | 10.0 | 8.0 | 8.0 | 7.7 |
|  | Six-month Libor dollars[b] ||||||||
| All countries | 7.74 | 6.25 | 6.54 | 9.48 | 12.12 | 14.35 | 13.42 | 16.70 |

*Source*: World Bank, Debtor Reporting System. Computed on an announcement basis.
[a] First half of year.
[b] London interbank offered late.

period 1975–76. But terms softened considerably in 1977, and an easier trend continued into 1981.

As opposed to syndicated loans, the external bond markets have not provided any significant amount of finance to developing countries. But in periods when industrial countries borrowed through bonds, developing countries could secure more financing through other channels. In effect, loanable funds in the private banking system have been a residual for the developing countries, depending on funds on deposit from the surplus nations and demand from the OECD nations. The effect of these influences was that flows of funds from both bank and bonds increased from $11 billion in 1973 to $52 billion in 1980.

The outlook for future flows of funds to the developing countries from the international banking system is uncertain. On the one hand, even though external debt has reached unprecedented levels for many countries, there is no danger of a complete cutoff of bank loans, and banks have rescheduled debt repayments in many cases where the borrower ran into financial difficulties. But they will exercise caution in these cases in the future, unlike the mid-and late seventies, when they provided financing merely for the asking. Euromarket lending has grown steadily at an annual rate of around 20 percent between 1978 and 1981, and is becoming an acceptable source of finance. Institutions in oil-importing developing countries have developed skills at borrowing when the market is favorable, instead of making frantic efforts when their needs are high. But the poorest nations continue to remain outsiders in this game. The two factors that will determine the outlook for the future are OPEC surpluses and the demand for funds of the industrial nations.

The likely magnitudes of OPEC surplus funds available through the year 2000 are discussed in the next chapter, but some structural changes currently taking place in international banking deserve mention. Arab banking has undergone rapid expansion since 1973, and indications are that a much larger share of surpluses in the Middle East nations will be handled by local financial institutions. With the exception of Kuwait, Arab investors have generally been conservative in their portfolio choices and have generally invested in government securities in the industrial nations. None of them, including Kuwaiti institutions, are likely to provide large-scale finance for the developing countries. Most funds invested in OECD countries, except those in equities, find outlets into the European banking system. This trend will continue unless the Middle East banks move into greater direct equity participation. Consequently, private financing for the developing countries will probably continue to come mainly from the European banking system. However, with higher levels of debt and greater risks of default, most lending would take place with higher spreads.

In conclusion, in the seventies capital flows from the private banking system played an important role in meeting the financial needs of developing countries in the seventies, enabling them to cope with higher deficits induced by oil price increases and yet maintain satisfactory rates of growth in national income. But structural changes have not taken place, which might allow these countries to wipe out recurring deficits and discharge their debt obligations at the same time, and the situation only worsened after the 1979–80 oil price increase. It is clear, therefore, that while private lending has a useful role to play, it is not a perfect substitute for concessional financing, which is more important in the difficult conditions of the eighties than it has been earlier.

## *Developing Energy in the Third World*

Financing of current-account deficits is of critical importance in ensuring a smooth functioning of the economic systems of most developing nations, which are confronted with a sudden rise in import costs or a sudden decline in export revenues. But from the global energy perspective, a transition from oil imports to domestic substitutes is desirable, and this calls for massive investments over the next few decades. A large number of developing nations have been constrained in pursuing such a transition for a lack of financial and other resources. Of considerable value to future global trends in energy consumption and production, therefore, is the availability of finance for energy developments in the Third World. To a great extent, of course, financing for energy developments cannot be separated from financing for other purposes, but there are some special features in energy development that merit separate treatment from the point of view of matching capital flows.

The World Bank has done an enormous amount of work in assessing the financial needs of the developing countries specifically for energy development (see World Bank 1980). The Brookhaven National Laboratory (1978) has also assessed the financial needs of developing countries for energy development. However, forecasting energy demand in the developing countries is fraught with uncertainty. Consequently, not only do forecasts of energy demand and supply vary from one study to another, there is little understanding of and agreement on the factors that influence the growth of demand. This results mainly from a paucity of empirical work in the energy sector in developing nations and from rule-of-thumb methods (such as were used by power companies in the United States until the midseventies). When national projections are aggregated, errors increase correspondingly. There is also the problem of forecasting variables such as energy prices and economic growth rates. Necessarily, therefore, projections of energy demand and supply can be discussed only in tentative terms and cannot provide precise estimates of the financing required.

In order to assess priorities and demand for financing, the World Bank classifies developing countries into two categories, oil exporters and oil importers. Oil exporters are divided into OPEC members and non-OPEC countries. Oil importers are broken down by oil imports as a percentage of commercial energy demand. Within these groupings there is further subdivision identifying nations with fuelwood problems. The large majority of countries fall within this subdivision, and certainly the largest population in the Third World is concentrated in countries within this subcategory. Table 8.4 shows the numbers of countries and their populations within each category.

The dominant variable in the World Bank's grouping is the dependence on imported oil (including that of oil-exporting nations), but other variables are also important in considering policies and priorities for energy-related financing. One is a country's overall dependence on imported energy. Based on current plans, a number of developing nations will become significant importers of steam coal,

Agenda for Change / 139

TABLE 8.4

Classification of Developing Countries, by Oil Imports and Fuelwood Scarcity

| Country Group | Countries without Severe Fuelwood Problems Number | Population[a] | Countries Lacking Fuelwood Number | Population[a] |
|---|---|---|---|---|
| Oil-exporting developing countries | 22 | 212.2 | 9 | 1,305.8 |
| OPEC | 11 | 95.1 | 3 | 233.6 |
| Non-OPEC | 11 | 117.1 | 6 | 1,072.2 |
| Oil-importing developing countries[b] | 34 | 522.2 | 58 | 1,274.5 |
| 0–25 | 5 | 121.5 | 3 | 719.2 |
| 26–50 | 3 | 34.6 | 5 | 184.4 |
| 51–75 | 5 | 203.9 | 5 | 41.5 |
| 76–100 | 21 | 62.2 | 45 | 329.4 |

*Source*: Compiled by author based on World Bank 1980, 1981*b*.
[a] Millions.
[b] Oil imports as percent of commercial energy reward.

notably the Republic of Korea, Taiwan, and the Philippines. These countries will require major investments in coal-related infrastructure, transport and handling facilities, and in some cases conversion of oil-burning to coal-burning equipment. Further, with higher coal imports, their economies will be vulnerable to higher coal prices, even though these are unlikely to change in a manner similar to oil prices. Another variable is a country's efficient use of energy. Significant improvements in efficient energy use are possible in many developing countries, depending on the structure of each economy and the pattern of energy consumption. Classification by intensity of energy use as well as by sector would provide a useful basis for international comparisons. Large amounts of financing would be required for modification and replacement of technology and equipment to improve energy efficiency.

Access to external financing is the third variable. As we have seen, financing by private banks varies considerably across countries. For countries that have had to rely on concessional funds, it is unlikely that private financing will be available in any significant measure. Other institutions would have to fill the gap between funds required for energy development and those available from domestic resources and official development assistance, especially for low-income countries with significant potential for development of indigenous energy resources. The availability of these resources is the final variable. A number of Third World nations have energy resources that have not been adequately explored and estimated. Any financing of energy development would, therefore, have to include preinvestment and exploration work, on the basis of which further decisions to invest could be made.

Even though various international organizations have stepped up their research on these variables, not enough is known for a precise country-by-country estimate of energy financing needs. This gap in information can be met through a global effort to explore and map energy resources in the relatively unexplored regions of the developing world. Even if they have the financing available, developing countries often shy away from investments in such research on account of the financial risks involved. Therefore, the prospects for a major increase in such investments, given the record of the past nine years, is limited unless a large part of the risk is assumed by external financing organizations, even beyond the stage of exploration.

The financing of petroleum development is an extreme case of risky investments. In the developing countries, the present institutional arrangements severely restrict exploration and, even where oil has been discovered, various constraints prevent development. Some useful work in assessing these problems has been done at the Massachusetts Institute of Technology, and in a summary paper, Tamir Agmon, Donald Lessard, and James Paddock (1981) establish the case for concessional financing in this field. They view the problem in terms of three components: the time lag in return on investment, the uncertainty of oil prices, and the need for external financing. There is generally a long lag between oil production and oil discovery. Returns on investments, therefore, normally occur in the distant future, while the development of the oilfield requires large investments in the interim. For a variety of reasons, largely political, producers do not normally shift oil-price risks to consumers. The uncertainty of oil prices also introduces uncertainty in future earnings. Investments become less attractive on account of this uncertainty. The authors show that the value of oil investments is more variable for large initial investments, as measured by the coefficient of variation of net present value. Further, if a country borrows in the international market to finance oil development, the variability of revenues will be even greater (since revenues depend on oil prices, but debt payments are fixed and invariable). This results in reduced investment and production in the countries developing their oil potential, a situation that can only be changed with external financing, which would transfer some of the oil-related risk. Most developing countries that discover substantial oil reserves will have to rely on external financing to develop these reserves. Riskiness of investment decisions increases with external credit. Equity investments or production sharing transfer some of these risks but are difficult to arrange on account of legal-contractual and political problems.

The authors examine the merits and shortcomings of various financing arrangements, including oil-linked bonds (such as those used by Mexico), lending by commercial banks, and official loans from the World Bank and the International Monetary Fund, and conclude that institutions such as the latter can play an important role. They could either assume a major part of the risks or not, depending on their ability to bear them. Further, the authors suggest the establishment of a formal energy development fund, on the rationale that industrialized oil-importing countries should provide a subsidy or transfer payment to fringe producer countries so that they may increase their investments in oil output capacity. Such investments would be a public good, because if they increased production they would reduce world oil prices. This recommendation is of far-reaching importance and can be extended from its limited purpose of increasing global oil output to increasing production of other forms of energy and to conserving oil through substitution of other fuels.

In essence, the recommendation of the Brandt Commission for a-dollar-a-barrel tax on oil imported by the developed countries for disbursement to the developing countries is also rooted in the public good concept. At current prices and import quantities, such a tax would result in a total annual funding of five billion dollars, which could be directed to energy developments in the Third World. But despite the likely mutual benefits, the industrial nations are hardly in a mood to agree on any significant increase in income transfers to the Third World. Besides, institutions already engaged in funding energy projects do not favor the idea of a new development fund for energy. The World Bank has been attempting to establish an energy affiliate for financing and overseeing a large energy program in the Third World.[2] But even though at one stage this arrangement—and a much expanded role for the Bank—seemed a virtual certainty, for the present it is not possible due to changes in the World Bank itself and to the Reagan administration's stand against large-scale multilateral financing for development in the Third World. The Bank's energy affiliate proposal has some appeal in that it emphasizes energy developments, which have been largely missing in its programs and activities in the past. The disadvantages of a mammoth centralized funding program of the nature proposed may outweigh the benefits. An aggregated, budget-bound scheme might lead to funding of second-best projects and programs. Additionally, a single

---

2. For a longer discussion of this proposal and its merits and demerits, see R. K. Pachauri 1982.

centralized institution operating within rigid policies and priorities might inhibit the development of expertise at the national level in nations with no history of formal energy policies and development. Energy is an integral part of all the economic activities of a nation, which can hardly be monitored by an international organization. A decentralized approach would be more flexible and effective, allowing national governments the freedom to decide on projects. For example, solar energy may be very attractive today and may seem to justify large-scale funding, but it could lose its luster tomorrow in the face of costs and technical inefficiency.

The most important factor that could diminish the effectiveness of the World Bank's proposed energy affiliate is the Bank's current inability to marshal resources without cutting into other funding. Any such venture must receive widespread support from its benefactors as well as its beneficiaries, but this has not materialized with the energy fund, particularly with respect to the United States and OPEC. Whereas the United States has withheld its support in a forthright and clear fashion,[3] OPEC has been much more vague in its obvious disapproval. In an OPEC-sponsored seminar in November 1981, Sheikh Yamani reportedly took the position that an energy affiliate of the type proposed by the World Bank would be too narrow and that a broader body, incorporating wider developmental considerations, would be preferable (Turner 1982, 161). It can be inferred from various reports that OPEC's support of the energy affiliate has been at best weak, if not entirely absent. Without doubt, OPEC does not want to be party to the birth of an institution that would rely heavily on it for support. Yet no global initiative in energy developments in the Third World requiring large-scale financing is possible without the explicit and generous support of OPEC. The industrial nations, assailed by insular tendencies, lower rates of growth, and recurrent stagflation, will probably not assume leadership in any such venture in the near future. They might join in a partnership if OPEC were to take the lead, but OPEC's current problems discount any such possibility.

## OPEC as a Source of Development Finance

Since 1973 and the accumulation of large surpluses in several OPEC countries, both North and South have eagerly monitored OPEC's assistance to other developing nations. Table 8.5 shows OPEC's performance as compared to that of the OECD countries. OPEC's assistance as a percentage of its gross national product declined

3. See "Energy Problems of the Third World," *Petroleum Economist* 53(1), 418–19.

TABLE 8.5

Official Assistance to Developing Countries, OPEC, OAPEC, and OECD Countries, 1960–1982 (percent of gross national product)

| OECD | 1960 | 1965 | 1970 | 1975 | 1978 | 1979 | 1980 | 1981 | 1982 |
|---|---|---|---|---|---|---|---|---|---|
| Italy | .03 | .04 | .06 | .01 | .01 | .01 | .01 | .02 | .04 |
| New Zealand | — | — | — | .14 | .03 | .01 | .01 | .01 | .00 |
| United Kingdom | .22 | .23 | .15 | .11 | .15 | .15 | .12 | .13 | .07 |
| Austria | — | .06 | .05 | .02 | .01 | .02 | .03 | .03 | .01 |
| Japan | .12 | .13 | .11 | .08 | .07 | .08 | .07 | .06 | .11 |
| Belgium | .27 | .56 | .30 | .31 | .23 | .27 | .24 | .25 | .21 |
| Finland | — | — | — | .06 | .04 | .06 | .08 | .09 | .08 |
| Netherlands | .19 | .08 | .24 | .24 | .34 | .26 | .32 | .37 | .29 |
| Australia | — | .08 | .09 | .10 | .08 | .06 | .04 | .06 | .08 |
| Canada | .11 | .10 | .22 | .24 | .17 | .13 | .11 | .12 | .13 |
| France | .01 | .12 | .09 | .10 | .08 | .07 | .08 | .11 | .10 |
| Germany, Fed. Rep. | .13 | .14 | .10 | .12 | .07 | .09 | .09 | .11 | .13 |
| Denmark | — | .02 | .10 | .20 | .21 | .28 | .28 | .20 | .22 |
| United States | .22 | .26 | .14 | .08 | .04 | .03 | .03 | .03 | .03 |
| Sweden | .01 | .07 | .12 | .41 | .37 | .41 | .34 | .31 | .36 |
| Norway | .02 | .04 | .12 | .25 | .39 | .33 | .28 | .25 | .33 |
| Switzerland | — | .02 | .05 | .10 | .08 | .06 | .08 | .07 | .09 |
| Total | .18 | .20 | .13 | .11 | .09 | .08 | .07 | .08 | .08 |

| OPEC | 1975 | 1976 | 1977 | 1978 | 1979 | 1980 | 1981 | 1982 |
|---|---|---|---|---|---|---|---|---|
| Nigeria | .04 | .19 | .10 | .05 | .04 | .04 | .20 | .08 |
| Algeria | .28 | .33 | .21 | .16 | .92 | .26 | .24 | .29 |
| Venezuela | .11 | .34 | .07 | .22 | .22 | .21 | .10 | .32 |
| Iran, Islamic Rep. | 1.12 | 1.16 | .22 | .33 | — | — | — | — |
| Iraq | 1.62 | 1.44 | .33 | .76 | 2.53 | 2.39 | .40 | — |
| Libya | 2.29 | .63 | .57 | .79 | .43 | 1.18 | 1.11 | .18 |
| Saudi Arabia | 7.76 | 6.46 | 5.24 | 8.39 | 5.55 | 5.09 | 3.58 | 2.82 |
| Kuwait | 7.40 | 3.63 | 8.10 | 5.46 | 3.50 | 3.40 | 3.55 | 4.86 |
| United Arab Emirates | 11.68 | 8.88 | 7.23 | 6.35 | 5.09 | 3.30 | 2.88 | 2.06 |
| Qatar | 15.58 | 7.95 | 7.56 | 3.38 | 6.18 | 4.03 | 3.75 | 3.80 |
| Total OAPEC | 5.73 | 4.23 | 3.95 | 4.69 | 3.54 | 3.44 | 2.87 | 2.42 |
| Total OPEC | 2.92 | 2.32 | 1.96 | 2.47 | 1.86 | 2.21 | 1.93 | 1.65 |

Source: World Bank, *World Development Report, 1985* (Washington, D.C., 1985), table 18.

significantly over the period 1975–80, even though it remained consistently higher than the percentage for the OECD nations. Within OPEC, the OAPEC members have been the highest contributors to development assistance, but their assistance, too, rapidly declined between 1975 and 1980. These declines reflect the rapid reduction in

TABLE 8.6
Institutions Distributing OPEC Aid

| Institution and Year Established[a] | Authorized/Subscribed Capital Billions of Dollars[b] | Percent Provided by OPEC | Geographical Mandate | Sectoral Priorities | Aid Commitments, 1973–80 Billions of Dollars | Percent in Grants |
|---|---|---|---|---|---|---|
| OF 1976 | 4.0[c] | 100 | Non-OPEC developing countries | Infrastructure/agriculture | 1.2[d] | 61 |
| KFAED 1961 | 3.6 | 100 | Developing countries | Infrastructure | 1.7 | 47 |
| SFD 1974 | 2.9 | 100 | Developing countries | Infrastructure | 2.3 | 49 |
| IsDB 1975 | 2.6 | 82 | Islamic countries | Infrastructure/industry | 0.1 | 50 |
| AFESD 1972 | 1.4 | 77 | Arab countries | Infrastructure | 1.0 | 32 |
| BADEA 1974 | 0.7 | 76 | Non-Arab Africa | Agriculture | 0.2 | 42 |
| AAAID 1977 | 0.5 | 87 | Arab countries | Agriculture | — | — |
| ADFAED 1971 | 0.5 | 100 | Developing countries except Latin America | Industry | 0.6 | 36 |
| IFED 1974 | 0.3 | 100 | Developing countries | — | 0.7[e] | 45 |
| ISF 1974 | 0.3 | 93 | Islamic countries | Education/relief aid | — | — |
| AFTAAC 1975 | 0.3 | 96 | Arab and African countries | Education/technical assistance | — | — |
| GAGSA 1953 | — | 100 | Arab countries | Education/health | 0.9[f] | 100 |
| IFAD 1976 | 1.0 | 43 | Member developing countries | Agriculture | 0.1 | — |

Source: Seymour 1979, 255–56; Kuwait Fund for Arab Economic Development; Chase World Information Corp.; OECD; Middle East Economic Survey.

[a] The following institutions either terminated operations or merged with other institutions: Gulf organisation for the Development of Egypt, Special Arab Aid Fund for Africa, and OAPEC special Account.
[b] As of May 1979.
[c] As of May 1980.
[d] To December 1979.
[e] To July 1979.
[f] To December 1976.

TABLE 8.6 (continued)

Abbreviations:

| | |
|---|---|
| OF: | The OPEC Fund for International Development |
| KFAED: | Kuwait Fund for Arab Economic Development |
| SFD: | Saudi Fund for Development |
| IsDB: | Islamic Development Bank |
| AFESD: | Arab Fund for Economic and Social Development |
| BADEA: | Arab Bank for Economic Development In Africa |
| AAAID: | Arab Authority for Agricultural Investment and Development |
| ADFAED: | Abu Dhabi Fund for Arab Economic Development |
| IFED: | Iraqi Fund for External Development |
| ISF: | Islamic Solidarity Fund |
| AFTAAC: | Arab Fund for Technical Assistance to Arab and African Countries |
| GAGSA: | General Authority for the Gulf and Southern Arabia |
| IFAD: | International Fund for Agricultural Development |

current-account surpluses between 1975 and 1978 and an obvious inertia during 1979 and 1980, years of high surpluses.

The impact of official assistance from OPEC can be better assessed by looking at the beneficiaries of such aid. Among many recent works dealing with this subject, Andre Simmons' (1981) book documents the record of Arab foreign aid, the dominant component of OPEC foreign assistance. In percentage terms, Arab aid has been much higher than OECD assistance, but from 1973 through 1977, 73 percent went to other Arab countries and 27 percent to non-Arab countries (19 percent to Asia, 8 percent to sub-Saharan Africa; see Simmons pages 26–27). Of the aid for non-Arab nations, the largest share went to the Muslim nations in Africa and Asia. Among the Arab recipients, Egypt, Syria, and Jordan received almost 90 percent of the aid. Despite pronouncements to the contrary and a genuine but small altruistic component, Arab aid is determined by political considerations. These political considerations are not so much a specific bias as a response to political pressure based on Islamic brotherhood. OPEC aid could be distributed more equitably within its existing charter through programs already in operation. These are shown in table 8.6.

A difference is demonstrated by some researchers between aid from the industrial nations and OPEC. (See Simmons 1981, 28–29, and Seymour 1980, 239–41.) Whereas transfers from industrial nations are generally recycled back to the donor nations through purchase of goods and services, OPEC aid represents a net transfer of financial resources. It is not recycled back to the donor nations, since the members of OPEC are generally not in a position to export goods and services other than oil and oil products. In fact, it is contended that Arab aid actually results in some benefit to the OECD nations, since the recipients generally spend such funds for goods and services from the industrial nations. But this argument can hardly be used against higher assistance, since OPEC itself spends the major share of its absorbable surpluses in the OECD nations, not in other developing countries.

Whereas OPEC aid and trading has provided little relief or benefit to most Third World nations, another channel for monetary flows, namely that of migrant worker remittances, has been of significance to a number of countries in recent years. The volume of such remittances as a ratio of merchandise exports is presented in table 8.7. It can be seen from these figures that the growing prosperity of the oil-exporting nations of the Middle East has provided a valuable opportunity for export of human capital. But there is some evidence that the increase in remittances observed in the late seventies levelled off in the early eighties. This trend is no doubt a result of the reduction in current-account surpluses and consequent scaling down of developmental expenditures. (Most of the migrant workers in the Middle East OPEC nations are employed as construction labor; hotel clerks, roomboys, and shopclerks; engineers; managers in financial institutions, business firms, and the oil industry; doctors and nurses; teachers; traders; and domestic help in Arab households. By far the greatest variation occurred in the construction industry, which experienced a major boom in the late seventies but has been declining in most of these nations in the early eighties.)

An important aspect of remittances from migrant labor is that they are not necessarily productive investments. A large proportion is

TABLE 8.7

Remittances of Migrant Workers in Middle East Countries to Their Home Country, as Ratio of Home Country Merchandise Exports

| Country | 1967 | 1973 | 1978–79 |
| --- | --- | --- | --- |
| Egypt | 0.044 | 0.117 | 0.888 |
| Syria | 0.032 | 0.104 | 0.088 |
| Southern Yemen | 0.837 | 1.340 | 5.638 |
| Yemen Arab Republic | — | 13.737 | 70.913 |
| Jordan | 0.583 | 0.608 | 1.754 |
| Sudan | 0.005 | 0.012 | 0.122 |
| India | 0.098 | 0.061 | 0.150 |
| Pakistan | — | 0.208 | 0.765 |
| Bangladesh | — | 0.049 | 0.210 |
| Republic of Korea | 0.104 | 0.008 | 0.007 |

Source: World Bank, *International Migrant Workers' Remittances: Issues and Prospects* (Washington, D. C.: 1981), table 3.

invested in real estate, agricultural land, and housing. The result of these preferences has been that land and real estate prices in some regions (in parts of Pakistan and in Kerala in India, for example) have escalated rapidly. The presence of a large temporary immigrant population causes social problems for the host country as well. Immigrants are generally single males, but eventually many are joined by females and children. Then follow demands by these workers for infrastructure, social services, and other facilities. These problems may diminish with the development of local skills and growth of indigenous populations. But nations like Kuwait, the United Arab Emirates, Qatar, and Saudi Arabia, which are highly dependent on foreign labor, would be forced to slow development if they reduced immigration. Their choice will critically affect their absorptive capacities, their oil production and export decisions, and the future of global energy supplies. I explore these and related issues within the framework of possible scenarios in the following chapter.

From the foregoing discussion, we can see that the nations of the Third World are likely to encounter more serious problems in the future than they have in the last ten years because of (1) higher prices for oil imports, which, combined with increased consumption, will increase their current-account deficits; (2) their inability to finance the development of indigenous energy; and (3) weak markets for exports of their manufactures, which will have to contend with stronger protectionist measures in the industrial nations and unstable and declining prices in world markets.

The North-South dialogue assumes some relevance on account of these likely developments, but in advocating change in the existing system, we must bear in mind the following realities. First, a new international economic order does not mean a revolution in the existing order. In a recent publication, Harold K. Jacobson and others (1983) studied the individuals involved in negotiations for a new economic order. On the basis of extensive structured interviews, the authors conclude that the negotiations are not really about the creation of a new order but a "struggle for the world product" (a phrase ascribed to Helmut Schmidt). It is, therefore, not surprising that the Brandt Commission report has largely fallen on deaf ears, given that it seems to the industrial nations as nothing more than the moralistic appeal of a trade union asking for a greater share of returns from production.

The prospects are hardly bright at present for any concensus on large-scale reform of the international monetary system, and even less so for any increased transfer of resources from North to South. The debate is likely to continue as a ritual without end. The final communique of the Williamsburg summit in mid-1983 deftly bypassed demands for a new economic order, leaving the impression that economic recovery in the industrial nations would bring about economic recovery and redressal of iniquities in the Third World. On the other hand, the policy document produced by the United Nations Conference on Trade and Development of 1983 listed a number of measures necessary for recovery and growth in the Third World, essentially the positions of the South. The impasse in negotiations also led the seventh summit of nonaligned countries held at New Delhi in March 1983 to suggest that a few government leaders from both North and South hold discussions on launching global negotiations. Quite understandably, the response from the North to this call has been mute. Given the recessionary conditions in the industrial nations and the domestic political compulsions of not conceding any real transfer of resources, the North is hardly going to agree to genuine negotiations. Consequently, it is in the interests of the South to concentrate on those areas of change where conflicts are not likely to arise, while maintaining the postures of the past ten years in discussions and negotiations. One such area is greater cooperation among countries of the South, which can serve as a complement to North-South relations. For importers of manufactured goods, such as OPEC countries, any increase in the products or technologies available from other developing countries would be desirable. A number of nations of the South have already reached such industrial maturity. In addition, there has been a sizable growth in multinational enterprises in the Third World, allowing them to offer other developing countries technologies, terms, and financial arrangements which may be more suitable than those from the North.

In essence, while it is undoubtedly difficult to convince nations to share part of their prosperity with those who are not so prosperous, it is not so difficult to drive home the point that stagnation in the poorest nations is detrimental to global peace, progress, and stability. There are numerous benefits to the North from greater and more rapid progress in the South. The dispute is only over what price the North is willing to pay for the progress of the South, and whether any concessions today would only open the floodgates for greater demands tomorrow. Clearly, no agenda for change can be implemented in an atmosphere of suspicion and misgivings, and hence what can be advocated, when all is said and done, is a long and persevering dialogue, in which substance should first give way to establishing a proper spirit.

CHAPTER 9

# FUTURE SCENARIOS AND PERSPECTIVES

Developments linking the aggregate demand for energy, supply and demand in the global oil market, oil exports and economic growth among the Middle East members of OPEC, and the North-South dialogue have been of paramount importance in global economic activity during the past decade. These linkages are likely to determine the movements and magnitudes of variables in the future. Analysts in the past have often followed a segmented approach, concentrating on some partial aspects of the total system. Consequently, predictions made with great assurance have invariably gone wrong, whether they related to global energy demand, total production of oil, the impact of new energy technology, the disposition of current-account surpluses of oil-exporting countries, or economic growth in nations of both North and South. I emphasized earlier that forecasting the time path of any one or all of these variables is a hazardous venture, susceptible as it is to influences that do not lend themselves to easy prediction. Assessment of future developments could go seriously wrong if causes and effects within the entire system are not explicitly considered. An attempt has been made in the following pages to develop scenarios of globally important variables and trends and to evaluate their plausibility in the light of causes and effects working through the linkages mentioned above.

The blinkers-on scenario and its implications for global economic developments were discussed in chapter 3. This analysis relates to the oil revenues likely to be earned by the capital-surplus countries of the Middle East, the feasible levels of absorption of these revenues, the disposition of net surpluses, and their effects on world monetary flows and the economies of the exporting nations themselves. The output levels for 1980, 1985, 1990, and 2000 are specified in table 3.7 for members of OPEC. The other variable for estimating OPEC revenues is the price of oil, and this has been assumed to grow for different grades of oil under two sets of price assumptions, $x$ and $y$, $x$ representing gradual real price increases and *y* representing similar trends except for a sudden price increase of 25 percent between 1984 and 1985 (which essentially represents the appreciation of the U.S. dollar in international markets in this period), followed by constant real prices between 1985 and 1990. These assumptions are detailed in table 9.1.

TABLE 9.1

Oil Price Assumptions Underlying Scenarios for the Future, Middle East Capital Surplus OPEC Countries (percent)

| Period | Price Assumption *x* | Price Assumption *y* |
|---|---|---|
| 1982–83 | −15 | −15 |
| 1983–84 | 0 | 0 |
| 1984–85 | 1 | 25 |
| 1985–90 | 2 | 0 |
| 1990–2000 | 3 | 3 |

TABLE 9.2

Projected Oil Production, Consumption, and Exports for Capital Surplus Middle East OPEC Countries (millions of barrels a day)

| Country | 1980 | 1982 | 1985 | 1990 | 2000 |
|---|---|---|---|---|---|
| Iraq | | | | | |
| Production | 2.60 | 0.92 | 3.50 | 3.50 | 3.50 |
| Consumption | 0.22 | 0.27 | 0.30 | 0.45 | 0.80 |
| Exports | 2.38 | 0.65 | 3.20 | 3.05 | 2.70 |
| Kuwait | | | | | |
| Production | 1.40 | 0.83 | 1.00 | 1.00 | 1.20 |
| Consumption | 0.05 | 0.06 | 0.05 | 0.07 | 0.12 |
| Exports | 1.35 | 0.77 | 0.95 | 0.93 | 1.08 |
| Libya | | | | | |
| Production | 1.80 | 1.15 | 1.20 | 1.00 | 1.00 |
| Consumption | 0.09 | 0.10 | 0.14 | 0.26 | 0.45 |
| Exports | 1.71 | 1.05 | 1.06 | 0.74 | 0.55 |
| Qatar | | | | | |
| Production | 0.50 | 0.32 | 0.50 | 0.30 | 0.27 |
| Consumption | 0.02 | 0.02 | 0.03 | 0.04 | 0.07 |
| Exports | 0.48 | 0.30 | 0.47 | 0.26 | 0.20 |
| Saudi Arabia | | | | | |
| Production | 9.70 | 6.45 | 8.50 | 8.50 | 8.50 |
| Consumption | 0.33 | 0.40 | 0.40 | 0.65 | 1.20 |
| Exports | 9.37 | 6.05 | 8.10 | 7.85 | 7.30 |
| United Arab Emirates | | | | | |
| Production | 1.70 | 1.21 | 1.40 | 1.40 | 1.40 |
| Consumption | 0.06 | 0.07 | 0.05 | 0.08 | 0.15 |
| Exports | 1.64 | 1.14 | 1.35 | 1.32 | 1.25 |

It needs to be noted that the future of oil prices is very uncertain—there is hardly any consensus among analysts—and the two assumptions in table 9.1 only illustrate the sensitivity of the system to variations in oil prices.

Future oil revenues are computed as a function of prices as determined by (1) the growth rates mentioned above, (2) oil production in keeping with the blinkers-on scenario, and (3) domestic consumption of oil and exports, as shown in table 9.2. As a result of continuing efforts at economic diversification, the nonoil gross domestic product of some OPEC nations is projected to grow at high rates in the future. Since the major oil-exporting nations are located in the Middle East, and the largest share of total capital surplus is accounted for by a small number of nations, we have confined our analysis to Iraq, Kuwait, Libya, Qatar, Saudi Arabia, and the United Arab Emirates (see table 9.3). Future growth would depend on the development of the nonoil sectors of the economy of each country, and, based on the analysis in chapter 5, I project optimistic growth rates for these nations.

The generation of capital surplus was computed in three separate steps, First, the investments made in the economy of a nation are calculated as a function of the nonoil gross domestic product. The total absorption of revenues, however, consists of investment and consump-

TABLE 9.3

Nonoil Gross Domestic Product, Middle East Capital Surplus OPEC Countries, 1974–2000 (percent)

| Year | Iraq | Kuwait | Libya | Qatar | Saudi Arabia | United Arab Emirates |
|---|---|---|---|---|---|---|
| 1974 | 32.6 | 19.6 | 42.7 | 30.0 | 89.9 | 66.6 |
| 1975 | 26.6 | 29.8 | 22.0 | 30.0 | 70.2 | 60.0 |
| 1976 | 23.4 | 34.2 | 17.8 | 30.0 | 42.2 | 58.3 |
| 1977 | 15.8 | 19.7 | 16.0 | 30.0 | 30.9 | 47.0 |
| 1978 | 14.0 | 8.9 | 12.2 | 20.0 | 25.5 | 1.0 |
| 1979 | 20.0 | 20.1 | 10.6 | 10.0 | 19.2 | 14.9 |
| 1980 | 25.0 | 14.3 | 16.3 | 15.0 | 21.1 | 23.8 |
| 1980–82[a] | 20.0 | 19.0 | 15.0 | 21.0 | 25.0 | 20.0 |
| 1982–85[a] | 15.0 | 15.0 | 12.0 | 16.0 | 12.5 | 15.0 |
| 1985–90[a] | 10.0 | 9.0 | 8.0 | 12.0 | 12.0 | 10.0 |
| 1990–2000[a] | 8.0 | 6.0 | 6.0 | 7.0 | 8.0 | 7.0 |
| 1976–80[b] | 19.6 | 19.4 | 14.6 | 21.0 | 27.7 | 29.0 |
| 1976–80[c] | — | 13.4 | 13.1 | — | 14.8 | 22.8 |

Source: World Bank 1981c, table A.3.
[a] Projected, constant prices.
[b] Average growth rate in current prices.
[c] Average growth rate in constant prices.

tion, and the latter is assumed to be a function of the former. The difference between total revenues and total absorption is the projected annual surplus. Finally, the total capital surplus is derived by adding the annual surplus to investment income, which is some percentage of the total foreign assets of a country held at the beginning of a year. A liberal return of 10 percent a year on foreign assets is assumed in my analysis.

On this basis, computation of future values results in nonoil gross domestic product (in 1980 dollars) as shown in table 9.4. If we take the population projected by the World Bank for these countries in the year 2000, we get nonoil GDP per capita of $5,100 for Iraq (population 23 million); $25,000 for Kuwait (population 2 million); $11,200 for Libya (population 5 million); and $25,720 for Saudi Arabia (population 15 million). This, combined with the oil component, would make these nations by far the richest in the world and capable of sustaining these levels of prosperity for long periods. It seems improbable, however, that

TABLE 9.4

Blinkers-on Scenario, Nonoil Gross Domestic Product, Middle East Capital-Surplus OPEC Countries, 1985–2000 (millions of dollars)

| Year | Iraq | Kuwait | Libya | Qatar | Saudi Arabia | United Arab Emirates |
|---|---|---|---|---|---|---|
| 1985 | 32.9 | 18.1 | 21.2 | 3.4 | 101.4 | 28.9 |
| 1986 | 36.2 | 19.7 | 22.9 | 3.8 | 113.6 | 31.8 |
| 1987 | 39.8 | 21.5 | 24.7 | 4.3 | 127.2 | 35.1 |
| 1988 | 43.8 | 23.4 | 26.7 | 4.8 | 142.5 | 38.5 |
| 1989 | 48.2 | 25.5 | 28.8 | 5.3 | 159.6 | 42.3 |
| 1990 | 53.1 | 27.8 | 31.1 | 6.1 | 178.7 | 46.5 |
| 1991 | 57.2 | 29.5 | 33.0 | 6.4 | 193.1 | 49.8 |
| 1992 | 61.8 | 31.3 | 35.1 | 6.9 | 208.4 | 53.3 |
| 1993 | 66.7 | 33.2 | 37.1 | 7.3 | 225.1 | 57.0 |
| 1994 | 72.1 | 35.2 | 39.3 | 7.9 | 243.1 | 61.0 |
| 1995 | 77.9 | 37.3 | 41.7 | 8.4 | 262.6 | 65.3 |
| 1996 | 84.1 | 39.5 | 44.2 | 9.1 | 283.6 | 69.8 |
| 1997 | 90.8 | 41.9 | 46.8 | 9.6 | 306.3 | 74.7 |
| 1998 | 98.1 | 44.4 | 49.6 | 10.3 | 330.8 | 80.1 |
| 1999 | 105.9 | 47.1 | 52.6 | 11.0 | 357.2 | 85.6 |
| 2000 | 114.4 | 49.9 | 55.8 | 11.8 | 385.8 | 91.6 |
| Population (millions) | 23 | 2 | 5 | * | 15 | * |
| GDP per capita (thousands) | 5.1 | 25.0 | 11.2 | * | 25.72 | * |

Source: World Bank, World Development Report, 1983–84.
* Projections for population not available

*Future Scenarios and Perspectives* / 153

such developments will actually take place without serious social, political, and religious disruption.

Scenarios of projected current-account surpluses and related variables for these nations are shown in tables 9.5 through 9.10. Surpluses are developed from their links with oil revenues, investment income, and stock of overseas capital assets and have been projected under oil price assumption $y$. Reviewing the values thus obtained, we can see that Iraq will generate increasing current-account surpluses with a correspondingly healthy increase in its overseas capital assets, except in the

TABLE 9.5

Iraq: Blinkers-on Scenario, Price Assumption $y$, 1981–2000
(billions of 1980 dollars)

| Year | Oil Revenue[a] | Absorption[b] | Surplus Component[c] | Investment Income[d] | Foreign Capital Assets[e] | Total Capital Surplus[f] |
|---|---|---|---|---|---|---|
| 1981 | 13.21 | 15.2 | −1.99 | 0.80 | 8.0 | −1.19 |
| 1982 | 8.42 | 17.7 | −9.28 | 0.69 | 6.93 | −8.59 |
| 1983 | 12.11 | 19.7 | −7.59 | −0.08 | −0.80 | −7.67 |
| 1984 | 20.92 | 21.8 | −0.88 | −0.77 | −7.70 | −1.65 |
| 1985 | 44.05 | 23.9 | 20.15 | −0.92 | −9.20 | 19.23 |
| 1986 | 43.64 | 25.5 | 18.14 | 0.81 | 8.11 | 18.95 |
| 1987 | 43.23 | 26.6 | 16.63 | 2.51 | 25.16 | 19.15 |
| 1988 | 42.82 | 28.4 | 14.42 | 4.24 | 42.39 | 18.66 |
| 1989 | 42.40 | 30.3 | 12.10 | 5.19 | 59.18 | 18.02 |
| 1990 | 41.99 | 31.6 | 10.39 | 7.54 | 75.39 | 17.93 |
| 1991 | 42.68 | 32.9 | 9.78 | 9.15 | 91.52 | 18.93 |
| 1992 | 43.37 | 37.9 | 5.47 | 10.85 | 108.56 | 16.32 |
| 1993 | 44.07 | 35.1 | 8.97 | 12.32 | 123.25 | 21.29 |
| 1994 | 44.78 | 36.5 | 8.28 | 14.25 | 142.41 | 22.52 |
| 1995 | 45.48 | 38.0 | 7.48 | 16.26 | 162.68 | 23.74 |
| 1996 | 46.18 | 39.6 | 6.58 | 18.40 | 184.05 | 24.98 |
| 1997 | 47.57 | 41.2 | 6.37 | 20.65 | 206.50 | 27.00 |
| 1998 | 47.60 | 42.8 | 4.80 | 23.08 | 230.80 | 27.88 |
| 1999 | 48.66 | 44.5 | 4.16 | 25.59 | 255.89 | 29.75 |
| 2000 | 49.93 | 46.2 | 3.73 | 28.26 | 282.66 | 31.99 |

[a] Calculated on the basis of oil exports and prices projected in tables 3.7 and 9.1.
[b] Nonoil gross domestic product calculated on basis of growth rates in table 9.2; domestic investments computed on the basis of investment-to-nonoil GDP ratios, which are based on the historic mean of this variable for 1973–80, with a subsequent annual decline of 3 percent a year. Absorption per year computed using the investment/absorption ratio based on the mean for 1973–80, with an annual increase of 1 percent subsequently.
[c] Difference between oil revenues and absorption in each year.
[d] Computed annually, using the stock of foreign assets at the beginning of a year and an annual yield of 10 percent thereon.
[e] Computed on the basis of existing assets at the beginning of the year plus 90 percent of total capital surplus generated during a year. (to allow for foreign aid, grants, and other leakages).
[f] Sum of surplus component and investment income.

154 / *The Political Economy of Global Energy*

TABLE 9.6

Kuwait: Blinkers-on Scenario, Price Assumption y, 1981–2000
(billions of 1980 dollars)[a]

| Year | Oil Revenue | Absorption | Surplus Component | Investment Income | Foreign Capital Assets | Total Capital Surplus |
|---|---|---|---|---|---|---|
| 1981 | 10.61 | 6.5  | 4.11  | 7.60  | 76.0   | 11.71 |
| 1982 | 9.52  | 7.3  | 2.22  | 8.65  | 86.54  | 10.87 |
| 1983 | 8.65  | 8.2  | 0.45  | 9.63  | 96.32  | 10.08 |
| 1984 | 9.28  | 8.8  | 0.48  | 10.54 | 105.39 | 11.02 |
| 1985 | 12.52 | 9.9  | 2.62  | 11.53 | 115.30 | 14.15 |
| 1986 | 12.45 | 10.4 | 2.05  | 12.80 | 128.00 | 14.85 |
| 1987 | 12.39 | 10.7 | 1.69  | 14.14 | 141.36 | 15.83 |
| 1988 | 12.34 | 11.2 | 1.14  | 15.56 | 155.60 | 16.70 |
| 1989 | 12.29 | 11.8 | 0.49  | 17.06 | 170.60 | 17.55 |
| 1990 | 12.26 | 12.4 | −0.14 | 18.48 | 184.89 | 18.34 |
| 1991 | 12.76 | 12.7 | 0.06  | 20.12 | 201.20 | 20.18 |
| 1992 | 13.28 | 13.1 | 0.18  | 21.93 | 219.30 | 22.11 |
| 1993 | 13.88 | 13.3 | 0.58  | 23.92 | 239.20 | 24.50 |
| 1994 | 14.53 | 14.0 | 0.53  | 26.12 | 261.20 | 26.65 |
| 1995 | 15.13 | 13.7 | 1.43  | 28.52 | 285.20 | 29.95 |
| 1996 | 15.73 | 13.9 | 1.83  | 31.21 | 312.10 | 33.04 |
| 1997 | 16.37 | 14.2 | 2.17  | 34.18 | 341.80 | 36.35 |
| 1998 | 17.19 | 14.4 | 2.79  | 37.45 | 374.50 | 40.24 |
| 1999 | 18.05 | 14.6 | 3.45  | 41.07 | 410.70 | 44.52 |
| 2000 | 19.02 | 14.8 | 4.32  | 45.07 | 450.70 | 49.39 |

[a] For calculations, see notes to table 9.4

period through 1985. It is assumed that Iraq's output will go up to 3.5 million barrels a day, despite the continuation of the Iran-Iraq war into 1985. In the case of Kuwait, the policies of today are expected to continue, with steady volumes of oil exports and sound investment strategies and a corresponding buildup of overseas capital assets, reaching $450.7 billion by the year 2000. Libya, on the other hand, will be facing some strains in coming years, with a small decline in oil revenues in real terms and recurring current-account deficits in the face of increasing absorptive power. But throughout the period covered, its overseas assets will keep building, on account of positive total capital surpluses arising out of positive investment income. In the case of Qatar, on the other hand, declining oil revenues and increasing absorption will lead to mounting total deficits and result in borrowing from external sources in the period 1977–2000.

The scenario for Saudi Arabia underlines the effect of its moderated oil production policies on internal economic conditions. Its deliberate strategy of producing below physical capacity to keep the world oil

TABLE 9.7

Libya: Blinkers-on Scenario, Price Assumption y, 1981–2000
(billions of 1980 dollars)[a]

| Year | Oil Revenue | Absorption | Surplus Component | Investment Income | Foreign Capital Assets | Total Capital Surplus |
|---|---|---|---|---|---|---|
| 1981 | 17.41 | 10.2 | 7.21 | 4.50 | 45.00 | 11.71 |
| 1982 | 13.91 | 11.4 | 2.51 | 5.55 | 55.54 | 8.06 |
| 1983 | 11.92 | 12.1 | −0.18 | 6.28 | 62.79 | 6.10 |
| 1984 | 11.96 | 12.9 | −0.94 | 6.80 | 68.28 | 5.88 |
| 1985 | 15.00 | 14.1 | 0.90 | 7.36 | 73.58 | 8.26 |
| 1986 | 13.86 | 14.5 | −0.64 | 8.10 | 81.00 | 7.46 |
| 1987 | 12.87 | 15.2 | −2.33 | 8.77 | 87.70 | 6.44 |
| 1988 | 11.88 | 15.6 | −3.72 | 9.34 | 93.49 | 5.63 |
| 1989 | 11.04 | 16.4 | −5.36 | 9.85 | 98.55 | 4.50 |
| 1990 | 10.47 | 17.0 | −6.53 | 10.26 | 102.60 | 3.73 |
| 1991 | 10.35 | 17.1 | −6.75 | 10.59 | 105.90 | 3.84 |
| 1992 | 10.36 | 17.5 | −7.14 | 10.93 | 109.30 | 3.79 |
| 1993 | 10.36 | 17.8 | −7.44 | 11.27 | 112.70 | 3.83 |
| 1994 | 10.35 | 18.2 | −7.85 | 11.61 | 116.10 | 3.76 |
| 1995 | 10.33 | 18.2 | −7.87 | 11.95 | 119.50 | 4.08 |
| 1996 | 10.31 | 18.6 | −8.29 | 12.31 | 123.17 | 4.03 |
| 1997 | 10.27 | 19.1 | −8.83 | 12.68 | 126.80 | 3.85 |
| 1998 | 10.34 | 19.4 | −9.06 | 13.02 | 130.20 | 3.96 |
| 1999 | 10.23 | 19.7 | −9.47 | 13.37 | 133.77 | 3.90 |
| 2000 | 10.46 | 20.1 | −9.64 | 13.73 | 137.28 | 4.09 |

[a] For calculations, see notes to table 9.4.

market in equilibrium will result in growth of capital assets at rates much lower than witnessed in the period 1974 to 1980. In fact, the sensitivity and precision of the strategy detailed in this scenario is implied in the small values of total capital surplus generated in the period. But again, it must be emphasized that these scenarios are purely illustrative and do not represent predictions of what is actually to come. The scenario for the United Arab Emirates, similar to that of Qatar, shows increasing current-account deficits, with net borrowing of $15.5 billion in the year 2000.

The values generated for this group of nations as a whole are shown in table 9.11, which shows aggregates for both price assumptions, $x$ and $y$. These values show the extreme sensitivity of surpluses and overseas holdings to oil prices, because foreign capital assets in the year 2000, for instance, are 484.93 under assumption $x$, but jump to 1,165.54 under assumption $y$. Yet, in general, the price of oil in the year 2000 under assumption $y$ is about 10 percent higher in real terms than under $x$. In 1985, however, the price difference is around 20 percent. But these differences in economic variables between the two price assumptions

TABLE 9.8

Qatar: Blinkers-on Scenario, 1981–2000
(billions of 1980 dollars) [a]

| Year | Oil Revenue | Absorption | Surplus Component | Investment Income | Foreign Capital Assets | Total Capital Surplus |
|---|---|---|---|---|---|---|
| 1981 | 5.09 | 3.3 | 1.79 | 1.200 | 12.00 | 2.99 |
| 1982 | 3.94 | 4.1 | −0.16 | 1.470 | 14.70 | 1.31 |
| 1983 | 3.91 | 4.3 | −0.39 | 1.590 | 15.88 | 1.20 |
| 1984 | 4.47 | 5.1 | −0.63 | 1.690 | 16.96 | 1.06 |
| 1985 | 6.56 | 5.6 | 0.96 | 1.790 | 17.90 | 2.75 |
| 1986 | 5.72 | 5.8 | −0.08 | 2.030 | 20.37 | 1.96 |
| 1987 | 5.02 | 6.5 | −1.48 | 2.210 | 22.13 | 0.73 |
| 1988 | 4.47 | 5.9 | −1.43 | 2.270 | 22.78 | 0.85 |
| 1989 | 3.91 | 7.4 | −3.49 | 2.350 | 23.54 | −1.13 |
| 1990 | 3.63 | 8.0 | −4.37 | 2.220 | 22.50 | −2.10 |
| 1991 | 3.59 | 8.2 | −4.61 | 2.060 | 20.60 | −2.79 |
| 1992 | 3.55 | 8.4 | −4.85 | 1.800 | 18.08 | −3.04 |
| 1993 | 3.57 | 8.5 | −4.93 | 1.530 | 15.34 | −3.40 |
| 1994 | 3.58 | 8.7 | −5.12 | 1.230 | 12.30 | −3.89 |
| 1995 | 3.59 | 9.1 | −5.51 | 0.880 | 8.80 | −4.62 |
| 1996 | 3.60 | 9.5 | −5.90 | 0.460 | 4.64 | −5.43 |
| 1997 | 3.62 | 9.4 | −5.78 | −0.025 | −0.25 | −5.80 |
| 1998 | 3.64 | 9.7 | −6.06 | −0.540 | −5.47 | −6.60 |
| 1999 | 3.68 | 10.1 | −6.42 | −1.140 | −11.40 | −7.56 |
| 2000 | 3.75 | 10.5 | −6.75 | −1.820 | −18.20 | −8.57 |

[a] For calculations, see notes to table 8.4.

really emphasize the importance of cumulative effects over a period of time. The post-1980 glut in oil and slide in oil prices obviously have effects that extend into the future, and it is to be expected that OPEC will try every stratagem to raise prices once again, even at some risk of political isolation. The U.S. strategy of building strong bridges with the moderate Arab states perhaps recognizes the importance to the stability of oil supplies and prices of political relations between the OPEC core nations and the western world.

I have not touched directly on the effects of higher oil prices on the demand side. A simulation-based study by George Daly, James Griffin, and Henry Steele (1983) develops scenarios of OPEC instability largely on the argument of high oil price elasticity of demand and high oil price elasticities of supply for non-OPEC producers and the so-called OPEC output maximizers (or high absorbers). I am not in agreement with the authors of this study, mainly because they assume a uniformly high price elasticity of −0.73 for the period through 2000, and their assumptions on non-OPEC supply responses are purely price induced, without

TABLE 9.9

Saudi Arabia: Blinkers-on Scenario, Price Assumption y, 1981–2000
(billions of 1980 dollars) [a]

| Year | Oil Revenue | Absorption | Surplus Component | Investment Income | Foreign Capital Assets | Total Capital Surplus |
|---|---|---|---|---|---|---|
| 1981 | 74.55 | 69.6 | 4.95 | 19.60 | 196.00 | 24.550 |
| 1982 | 70.66 | 82.6 | −11.94 | 21.81 | 218.10 | 9.870 |
| 1983 | 66.12 | 90.8 | −24.68 | 22.69 | 226.98 | − 1.980 |
| 1984 | 72.97 | 90.8 | −17.83 | 22.52 | 225.19 | 4.690 |
| 1985 | 100.52 | 97.8 | 2.72 | 22.94 | 229.40 | 25.660 |
| 1986 | 99.90 | 102.2 | − 2.30 | 25.25 | 252.49 | 22.950 |
| 1987 | 99.28 | 108.5 | − 9.22 | 27.31 | 273.14 | 18.090 |
| 1988 | 98.66 | 108.7 | −10.04 | 28.94 | 289.40 | 18.900 |
| 1989 | 98.04 | 122.5 | −24.46 | 30.64 | 306.40 | 6.180 |
| 1990 | 97.42 | 128.4 | −30.98 | 31.19 | 311.90 | 0.216 |
| 1991 | 99.57 | 129.3 | −29.73 | 31.21 | 312.10 | 1.480 |
| 1992 | 101.77 | 133.5 | −31.73 | 31.34 | 313.40 | 0.380 |
| 1993 | 104.00 | 134.1 | −30.10 | 31.31 | 313.10 | 1.210 |
| 1994 | 106.40 | 138.2 | −31.80 | 31.41 | 314.17 | 0.400 |
| 1995 | 108.86 | 138.2 | −29.34 | 31.38 | 313.80 | 2.040 |
| 1996 | 111.38 | 140.0 | −28.62 | 31.56 | 315.60 | 2.940 |
| 1997 | 113.97 | 141.4 | −27.43 | 31.82 | 318.20 | 4.390 |
| 1998 | 116.30 | 144.6 | −28.30 | 32.21 | 322.10 | 3.910 |
| 1999 | 118.98 | 148.2 | −29.22 | 32.56 | 325.60 | 3.340 |
| 2000 | 121.55 | 151.5 | −29.95 | 32.86 | 328.60 | 2.910 |

[a] For calculations, see notes to table 9.4.

regard to political, financial, or technical constraints. I do not believe that the price elasticity of demand (unusually high in itself) will remain at this level once the initial phase of conservation, arising from modest investments and expenditures, is over. Further, the demand scenarios in this book implicitly assume a conservation response in keeping with the price assumptions, which are reasonably modest in comparison with the increase of the 1970s.

Warnings have been sounded by various researchers and commentators on the growth of petrodollar flows and deposits,[1] but little has been done to assess the quantitative dimensions of the problem. Given current patterns of OPEC investment behavior, institutional channels for investment flows, and recipient country attitudes, such a large transfer of asset ownership with reasonable rates of return to the investors appears highly unlikely. And without reasonable returns,

1. See, for instance, Normal Gall, "How Much More Can the System Take ?" *Forbes*, June 23, 1980.

## TABLE 9.10

United Arab Emirates: Blinkers-on Scenario, Price Assumption y, 1981–2000 (billions of 1980 dollars)

| Year | Oil Revenue | Absorption | Surplus Component | Investment Income | Foreign Capital Assets | Total Capital Surplus |
|---|---|---|---|---|---|---|
| 1981 | 15.25 | 14.4 | 0.85 | 4.20 | 42.00 | 5.05 |
| 1982 | 14.56 | 16.4 | −1.84 | 4.65 | 46.50 | 2.81 |
| 1983 | 13.03 | 17.9 | −4.87 | 4.90 | 49.00 | 0.03 |
| 1984 | 13.79 | 19.7 | −5.91 | 4.90 | 49.03 | −1.01 |
| 1985 | 18.33 | 21.4 | −3.07 | 4.84 | 48.40 | 1.77 |
| 1986 | 18.19 | 22.5 | −4.31 | 5.00 | 49.99 | 0.69 |
| 1987 | 18.05 | 23.8 | −5.75 | 5.06 | 50.60 | −0.68 |
| 1988 | 17.92 | 24.5 | −6.58 | 5.00 | 49.98 | −1.58 |
| 1989 | 17.92 | 25.7 | −7.78 | 4.85 | 48.55 | −2.92 |
| 1990 | 17.92 | 27.1 | −9.18 | 4.59 | 45.92 | −4.58 |
| 1991 | 18.35 | 27.2 | −8.85 | 4.18 | 41.79 | −4.67 |
| 1992 | 18.80 | 27.8 | −9.00 | 3.75 | 37.58 | −5.24 |
| 1993 | 19.26 | 28.7 | −9.44 | 3.28 | 32.86 | −6.15 |
| 1994 | 19.73 | 28.9 | −9.17 | 2.73 | 27.32 | −6.43 |
| 1995 | 20.21 | 29.2 | −8.99 | 2.15 | 21.52 | −6.84 |
| 1996 | 20.74 | 30.1 | −9.36 | 1.53 | 15.36 | −7.82 |
| 1997 | 21.20 | 30.3 | −9.10 | 0.83 | 8.32 | −8.26 |
| 1998 | 21.73 | 30.4 | −8.67 | 0.09 | 0.88 | −8.58 |
| 1999 | 22.28 | 31.3 | −9.02 | −0.68 | −6.80 | −9.70 |
| 2000 | 22.79 | 32.1 | −9.31 | −1.55 | −15.50 | −10.86 |

[a] For calculations, see notes to table 9.4.

## TABLE 9.11

Blinkers-on Scenario, Middle East Capital Surplus OPEC Countries, Selected Years, 1981–2000 (billions of 1980 dollars)

| Price Assumption[a] and Year | Oil Revenue | Foreign Capital Assets | Total Capital Surplus |
|---|---|---|---|
| Price assumption | | | |
| 1981 | 136.12 | 379.00 | 57.25 |
| 1985 | 159.13 | 475.39 | 33.97 |
| 1990 | 163.80 | 579.14 | −2.80 |
| 2000 | 203.13 | 484.93 | −23.22 |
| Price assumption | | | |
| 1981 | 136.12 | 379.00 | 54.82 |
| 1985 | 196.36 | 475.38 | 71.82 |
| 1990 | 183.69 | 743.20 | 33.53 |
| 2000 | 227.60 | 1165.54 | 68.95 |

[a] See table 9.1.

TABLE 9.12

Assumptions Underlying Scarcity Scenario, Middle East Capital Surplus OPEC Countries, Selected Years 1980–2000 (millions of barrels a day)

| Country | 1980 | 1982 | 1985 | 1990 | 2000 |
|---|---|---|---|---|---|
| Iraq | | | | | |
| Production | 2.60 | 0.92 | 3.00 | 3.00 | 2.00 |
| Consumption | 0.22 | 0.27 | 0.30 | 0.45 | 0.80 |
| Exports | 2.38 | 0.65 | 2.70 | 2.55 | 1.20 |
| Memorandum | | | | | |
| Annual Growth in nonoil GDP over 2 years (percent) | — | 20 | 10 | 8 | 8 |
| Kuwait | | | | | |
| Production | 1.40 | 0.83 | 0.80 | 0.60 | 0.50 |
| Consumption | 0.05 | 0.06 | 0.05 | 0.07 | 0.12 |
| Exports | 1.35 | 0.77 | 0.75 | 0.53 | 0.38 |
| Memorandum | | | | | |
| Annual Growth in nonoil GDP over 2 years (percent) | — | 19 | 10 | 7 | 6 |
| Libya | | | | | |
| Production | 1.80 | 1.15 | 0.80 | 0.70 | 0.50 |
| Consumption | 0.09 | 0.10 | 0.14 | 0.26 | 0.45 |
| Exports | 1.71 | 1.05 | 0.66 | 0.44 | 0.05 |
| Memorandum | | | | | |
| Annual Growth in nonoil GDP over 2 years (percent) | — | 15 | 8 | 6 | 5 |
| Qatar | | | | | |
| Production | 0.50 | 0.33 | 0.400 | 0.40 | 0.30 |
| Consumption | 0.02 | 0.02 | 0.025 | 0.04 | 0.07 |
| Exports | 0.48 | 0.30 | 0.375 | 0.36 | 0.23 |
| Memorandum | | | | | |
| Annual Growth in nonoil GDP over 2 years (percent) | — | 21 | 10 | 7 | 6 |
| Saudi Arabia | | | | | |
| Production | 9.70 | 6.45 | 7.00 | 7.50 | 7.00 |
| Consumption | 0.33 | 0.40 | 0.40 | 0.65 | 1.20 |
| Exports | 9.37 | 6.05 | 6.60 | 6.85 | 5.80 |
| Memorandum | | | | | |
| Annual Growth in nonoil GDP over 2 years (percent) | — | 25 | 11 | 6 | 6 |
| UAE | | | | | |
| Production | 1.70 | 1.21 | 1.00 | 0.80 | 0.60 |
| Consumption | 0.06 | 0.07 | 0.08 | 0.10 | 0.15 |
| Exports | 1.64 | 1.14 | 0.92 | 0.70 | 0.45 |
| Memorandum | | | | | |
| Annual Growth in nonoil GDP over 2 years (percent) | — | 20 | 10 | 6 | 6 |
| Total | | | | | |
| OPEC Production | | | 20.3 | 19.8 | 16.1 |

larger surpluses would not be viewed with favor by the capital-surplus oil exporters.

Oil production and export decisions by members of OPEC, and more particularly by the capital-surplus nations, cannot be based on profit-maximizing behavior, as shown earlier. These decisions are intimately connected with the developmental goals and economic performance of these nations. Before the oil price increases of the last decade, those nations that today have the largest oil reserves also happened to be among the least developed. Even today, outside the oil sector, these nations clearly show their history of economic weakness, which they are trying to rapidly overcome. Oil is regarded as the gift of God in bringing about an economic transformation in these lands and is, therefore, to be produced and exported in such quantities as would make a wise transformation feasible. The frenzied drive to invest in nonoil sectors is based on a recognition of the temporal nature of their oil wealth and the imperative of activating other forms of productive capacity. This rate of development, of course, is constrained by the revenues that can be generated as well as the country's absorptive capacity.

TABLE 9.13

Iraq: Scarcity Scenario, 1981–2000 (billions of 1980 dollars)[a]

| Year | Oil Revenue | Absorption | Surplus Component | Investment Income | Foreign Capital Assets | Total Capital Surplus |
|---|---|---|---|---|---|---|
| 1981 | 13.20 | 15.2 | − 2.00 | 0.800 | 8.00 | 1.20 |
| 1982 | 8.45 | 17.7 | − 9.25 | 0.900 | 9.08 | − 8.34 |
| 1983 | 11.56 | 19.1 | − 7.54 | 0.157 | 1.57 | − 7.38 |
| 1984 | 18.50 | 20.1 | − 1.60 | −0.500 | − 5.07 | − 2.10 |
| 1985 | 30.03 | 20.7 | 9.33 | −0.690 | − 6.96 | 8.64 |
| 1986 | 30.29 | 21.8 | 8.50 | 0.081 | 0.81 | 8.58 |
| 1987 | 30.54 | 22.4 | 8.14 | 0.850 | 8.50 | 8.99 |
| 1988 | 30.80 | 23.4 | 7.40 | 1.660 | 16.59 | 9.06 |
| 1989 | 31.05 | 24.7 | 6.35 | 2.470 | 24.74 | 8.82 |
| 1990 | 31.30 | 25.1 | 6.20 | 3.260 | 32.68 | 9.46 |
| 1991 | 29.84 | 26.2 | 3.64 | 4.120 | 41.20 | 7.76 |
| 1992 | 28.52 | 27.4 | 1.12 | 4.820 | 48.18 | 5.94 |
| 1993 | 27.36 | 27.9 | 0.54 | 5.350 | 53.52 | 5.90 |
| 1994 | 26.25 | 29.1 | − 2.85 | 5.880 | 58.83 | 3.03 |
| 1995 | 25.04 | 30.3 | − 5.26 | 6.150 | 61.55 | 0.89 |
| 1996 | 24.03 | 31.5 | − 7.47 | 6.230 | 62.35 | − 1.23 |
| 1997 | 22.94 | 32.9 | − 9.96 | 6.120 | 61.24 | − 3.84 |
| 1998 | 21.92 | 34.2 | −12.28 | 5.780 | 57.78 | − 6.50 |
| 1999 | 20.97 | 35.5 | −14.53 | 5.200 | 51.93 | − 9.33 |
| 2000 | 19.79 | 36.8 | −17.01 | 4.350 | 43.53 | −12.65 |

[a] For calculations, see notes to table 9.4.

A country's absorptive capacity is not easy to determine in isolation from the resources available for absorbing. To a large extent, the pattern of investment will be determined by the financial resources available—more money may lead only to more grandiose forms of investment. But based on the trends and the lessons of the seventies, one could come up with estimates of absorptive capacity in the future, at least for the capital-surplus nations, which are unlikely to face any lack of funds in the next two decades. But the initial euphoria of 1973–76 has been replaced by some caution, resulting from global inflation since 1973, the decline in surpluses in 1977–78, and the sharp reduction in demand since 1980. There is, therefore, evidence that a more mature phase of economic management has set in, even though the second round of oil price increases in 1979–80 brought about a splurge in spending similar to that of the midseventies.

Quite apart from the economic effects of an accelerated pace of investments, the social implications are even more serious. Most of these nations are sparsely populated and have a history of employing migrant labor. With a large increase in construction and other activities,

TABLE 9.14

Kuwait: Scarcity Scenario, 1981–2000 (billions of 1980 dollars)[a]

| Year | Oil Revenue | Absorption | Surplus Component | Investment Income | Foreign Capital Assets | Total Capital Surplus |
|---|---|---|---|---|---|---|
| 1981 | 10.61 | 6.5 | 4.110 | 7.60 | 76.00 | 11.71 |
| 1982 | 9.52 | 7.3 | 2.220 | 8.65 | 86.54 | 10.87 |
| 1983 | 8.03 | 7.7 | 0.330 | 9.63 | 96.32 | 9.96 |
| 1984 | 7.96 | 8.1 | −0.135 | 10.53 | 105.30 | 10.39 |
| 1985 | 7.99 | 8.6 | −0.610 | 11.46 | 114.60 | 10.85 |
| 1986 | 7.60 | 8.9 | −1.300 | 12.43 | 124.36 | 11.13 |
| 1987 | 7.20 | 9.0 | −1.800 | 13.43 | 134.37 | 11.63 |
| 1988 | 6.78 | 9.4 | −2.620 | 14.48 | 144.83 | 11.86 |
| 1989 | 6.45 | 9.6 | −3.150 | 15.55 | 155.50 | 12.40 |
| 1990 | 6.23 | 9.9 | −3.670 | 16.66 | 166.60 | 12.99 |
| 1991 | 6.17 | 10.2 | −4.030 | 17.83 | 178.29 | 13.80 |
| 1992 | 5.11 | 10.4 | −5.290 | 19.07 | 190.71 | 13.78 |
| 1993 | 6.03 | 10.7 | −4.670 | 20.31 | 203.10 | 15.64 |
| 1994 | 5.95 | 10.7 | −4.750 | 21.71 | 217.10 | 16.96 |
| 1995 | 5.99 | 10.9 | −4.910 | 23.24 | 232.37 | 18.33 |
| 1996 | 5.96 | 11.1 | −5.140 | 24.88 | 248.86 | 19.74 |
| 1997 | 5.93 | 11.3 | −5.370 | 26.66 | 266.60 | 21.29 |
| 1998 | 5.95 | 11.5 | −5.550 | 28.57 | 285.76 | 23.02 |
| 1999 | 5.98 | 11.6 | −5.620 | 30.65 | 306.47 | 25.03 |
| 2000 | 6.00 | 11.7 | −5.700 | 32.89 | 328.99 | 27.19 |

[a] For calculations, see notes to table 9.4.

the rate of increase in expatriate labor has gone up substantially, particularly in the United Arab Emirates, Kuwait, and Qatar, whose expatriate populations now outnumber the local citizenry. While the expatriates are allowed to perform the "dirty" jobs, some end up in the highest paying professions, since inadequate educational systems have not prepared enough citizens for these jobs.

In 1975 the expatriate population in the six capital surplus nations being discussed was around 2.5 million in a total population of some 20 million. About a million of these immigrants were dependents. These families live in poor conditions, alienated from the dominant culture. The problem is particularly serious in the United Arab Emirates, where in 1980 expatriates numbered almost three times the local population, with a preponderance of non-Arab immigrants with cultures, languages, and religions very different from those of the locals. The World Bank (1981c, 21) projects that with an average growth rate of 10 percent a year between 1975 and 1985, the expatriate work force may go from 1.5 million to about 3.5 million. With a corresponding growth in the dependent population, expatriates could number 11.0 million by 1985. (Unfortunately, data are not yet available to verify this.)

TABLE 9.15

Libya: Scarcity Scenario, 1981–2000 (billions of 1980 dollars)[a]

| Year | Oil Revenue | Absorption | Surplus Component | Investment Income | Foreign Capital Assets | Total Capital Surplus |
|---|---|---|---|---|---|---|
| 1981 | 17.41 | 10.2 | 7.21 | 4.50 | 45.00 | 11.71 |
| 1982 | 13.98 | 11.4 | 2.58 | 5.56 | 55.57 | 8.14 |
| 1983 | 10.18 | 11.7 | − 1.52 | 6.23 | 62.89 | 4.71 |
| 1984 | 8.72 | 12.1 | − 3.38 | 6.71 | 67.13 | 3.33 |
| 1985 | 7.55 | 12.6 | − 5.05 | 7.01 | 70.13 | 1.96 |
| 1986 | 7.11 | 12.6 | − 5.49 | 7.19 | 71.89 | 1.70 |
| 1987 | 6.66 | 13.0 | − 6.34 | 7.34 | 73.42 | 1.00 |
| 1988 | 6.31 | 13.2 | − 6.89 | 7.43 | 74.32 | 0.54 |
| 1989 | 5.94 | 13.5 | − 7.56 | 7.48 | 74.80 | − 0.08 |
| 1990 | 5.55 | 13.9 | − 8.35 | 7.47 | 74.72 | − 0.88 |
| 1991 | 4.55 | 13.8 | − 9.25 | 7.39 | 73.93 | − 1.86 |
| 1992 | 3.75 | 14.1 | −10.35 | 7.22 | 72.26 | − 3.13 |
| 1993 | 3.17 | 14.2 | −11.03 | 6.94 | 69.44 | − 4.09 |
| 1994 | 2.56 | 14.4 | −11.84 | 6.57 | 65.76 | − 5.27 |
| 1995 | 2.19 | 14.3 | −12.11 | 6.10 | 61.01 | − 6.01 |
| 1996 | 1.81 | 14.3 | −12.49 | 5.56 | 55.60 | − 6.93 |
| 1997 | 1.55 | 14.5 | −12.95 | 4.93 | 49.36 | − 8.01 |
| 1998 | 1.12 | 14.7 | −13.58 | 4.21 | 42.14 | − 9.37 |
| 1999 | 0.99 | 14.7 | −13.71 | 3.37 | 33.70 | −10.34 |
| 2000 | 0.85 | 14.8 | −13.95 | 2.44 | 24.40 | −11.51 |

[a] For calculations, see notes to table 9.4.

This would mean an annual growth rate of 13.5 percent as against 3.5 percent for the local population. Since all the countries in this group have commitments to social services and infrastructure, the cost of hiring expatriate labor is not confined to their wages alone but includes subsidies for food, petroleum products, and utilities, and which apply to the entire population, including immigrants. Governments are, therefore, likely to discriminate in favor of their citizens in various other ways, typically, by employing them in important and high-paying jobs instead of immigrants with superior qualifications. This trend would undermine efficiency, placing a premium on citizenship rather than merit, and would enhance social divisions over the long run.

Given the surplus of capital and scarcity of labor in these countries, there will be a greater effort to move into capital-intensive production. But such technologies have to be imported, increasing dependence on foreign suppliers, contractors, and technical personnel. Further, given the small populations of these nations, most goods and services produced would have to be sold in foreign markets. An example of this dependence on markets overseas is the petrochemical industry of

TABLE 9.16

Qatar: Scarcity Scenario, 1981–2000 (billions of 1980 dollars)[a]

| Year | Oil Revenue | Absorption | Surplus Component | Investment Income | Foreign Capital Assets | Total Capital Surplus |
|---|---|---|---|---|---|---|
| 1981 | 5.09 | 3.3 | 1.79 | 1.20 | 12.00 | 2.99 |
| 1982 | 3.99 | 4.1 | −0.11 | 1.47 | 14.70 | 1.36 |
| 1983 | 3.64 | 4.3 | −0.66 | 1.59 | 15.92 | 0.93 |
| 1984 | 3.91 | 4.5 | −0.59 | 1.67 | 16.76 | 1.08 |
| 1985 | 4.23 | 4.8 | −0.57 | 1.77 | 17.73 | 1.20 |
| 1986 | 4.27 | 4.7 | −0.43 | 1.88 | 18.81 | 1.45 |
| 1987 | 4.32 | 4.9 | −0.58 | 2.01 | 20.11 | 1.43 |
| 1988 | 4.37 | 5.1 | −0.73 | 2.14 | 21.40 | 1.41 |
| 1989 | 4.42 | 5.3 | −0.88 | 2.26 | 22.67 | 1.38 |
| 1990 | 4.48 | 5.3 | −0.82 | 2.39 | 23.91 | 1.57 |
| 1991 | 4.41 | 5.5 | −1.09 | 2.53 | 25.32 | 1.44 |
| 1992 | 4.34 | 5.7 | −1.36 | 2.66 | 26.60 | 1.30 |
| 1993 | 4.27 | 5.9 | −1.63 | 2.77 | 27.77 | 1.14 |
| 1994 | 4.20 | 5.8 | −1.60 | 2.88 | 28.79 | 1.28 |
| 1995 | 4.17 | 6.1 | −1.93 | 2.99 | 29.94 | 1.06 |
| 1996 | 4.07 | 6.2 | −2.13 | 3.09 | 30.89 | 0.96 |
| 1997 | 4.01 | 6.1 | −2.09 | 3.17 | 31.75 | 1.08 |
| 1998 | 3.94 | 6.3 | −2.36 | 3.27 | 32.72 | 0.91 |
| 1999 | 3.89 | 6.4 | −2.51 | 3.35 | 33.54 | 0.84 |
| 2000 | 3.84 | 6.4 | −2.56 | 3.43 | 34.29 | 0.87 |

[a] For calculations, see notes to table 9.4.

164 / *The Political Economy of Global Energy*

Kuwait, which, in the absence of adequate market outlets, has been producing at low capacity since its inception. Finally, a satisfactory performance of most capital-intensive industries depends on a developed infrastructure in other sectors of the economy, and these have a high dependence on skilled labor. Thus higher growth rates can be sustained only through larger inflows of immigrant labor. And even when construction activity ends, immigrant labor would be needed to man a number of operations taken for granted in a developed country.

Constraints in the availability of labor are therefore likely to be a major determinant of future growth strategies. Given this outlook, the absorptive capacity and investment levels characterizing the blinkers-on scenario appear much too optimistic. It is likely that capital surpluses and the growth of foreign assets will be actually higher if production and export levels are as high as assumed in the scenario. The implication is very strong that a much lower growth strategy will be followed by these countries in the aggregate. If this were to happen, then much lower targets will be set for oil production and exports, since the real returns

TABLE 9.17

Saudi Arabia: Scarcity Scenario, 1981–2000 (billions of 1980 dollars)[a]

| Year | Oil Revenue | Absorption | Surplus Component | Investment Income | Foreign Capital Assets | Total Capital Surplus |
|---|---|---|---|---|---|---|
| 1981 | 74.55 | 69.6 | 4.95 | 19.60 | 196.0 | 24.55 |
| 1982 | 70.66 | 82.6 | −11.94 | 21.81 | 218.1 | 9.87 |
| 1983 | 61.75 | 86.8 | −25.05 | 22.70 | 226.9 | − 2.35 |
| 1984 | 63.53 | 91.1 | −27.57 | 22.48 | 224.8 | − 5.10 |
| 1985 | 66.17 | 93.8 | −27.63 | 22.02 | 220.2 | − 5.61 |
| 1986 | 68.01 | 92.7 | −24.69 | 21.51 | 215.1 | − 3.17 |
| 1987 | 69.78 | 83.1 | −13.32 | 21.23 | 212.3 | 7.91 |
| 1988 | 71.71 | 92.7 | −20.99 | 21.94 | 219.4 | 0.95 |
| 1989 | 73.67 | 94.2 | −20.53 | 22.02 | 220.2 | 1.50 |
| 1990 | 75.81 | 93.3 | −17.49 | 22.15 | 221.5 | 4.66 |
| 1991 | 76.71 | 92.5 | −15.79 | 22.57 | 225.7 | 6.78 |
| 1992 | 77.61 | 93.6 | −15.99 | 23.18 | 231.8 | 7.19 |
| 1993 | 78.49 | 92.4 | −13.91 | 23.82 | 238.2 | 9.91 |
| 1994 | 79.48 | 93.4 | −13.92 | 24.71 | 247.1 | 10.79 |
| 1995 | 80.45 | 91.6 | −11.15 | 25.68 | 256.8 | 14.53 |
| 1996 | 81.40 | 92.4 | −11.00 | 26.99 | 269.9 | 15.99 |
| 1997 | 82.32 | 90.2 | − 7.88 | 28.43 | 284.3 | 20.55 |
| 1998 | 83.25 | 90.7 | − 7.45 | 30.28 | 302.8 | 22.83 |
| 1999 | 84.24 | 91.1 | − 6.86 | 32.33 | 323.3 | 25.47 |
| 2000 | 86.14 | 91.4 | − 5.26 | 34.62 | 346.2 | 29.36 |

[a] For calculations, see notes to table 9.4.

*Future Scenarios and Perspectives* / 165

from, and new opportunities for, investments overseas would decline as the size of surpluses to be invested decreases.

We now turn to the scarcity scenario, the underlying basis for which is quantified in table 9.12. This scenario rests essentially on a lower growth strategy in the Middle East capital-surplus OPEC nations, and, consequently, lower oil production and exports (see tables 9.13 through 9.18). For the sake of simplicity, we assume that prices remain the same as under the blinkers-on scenario. The values of capital assets accumulated and surpluses generated from year to year under the scarcity scenario are very different from those under the blinkers-on scenario. For instance, Iraq's growth in nonoil gross domestic product, despite lower values, results in lower or negative current-account surpluses and overseas capital assets. To a lesser degree this is true of Kuwait as well, but since the absorptive capacity of Kuwait is relatively low and the size of its oil revenues and capital assets relatively high, this country's overall position remains comfortable. But even though Libya is a low absorber, it does not fare as well as Kuwait, and under this scenario total capital deficits appear persistently in the 1990s. Qatar's

TABLE 9.18

United Arab Emirates: Scarcity Scenario, 1981–2000 (billions of 1980 dollars)[a]

| Year | Oil Revenue | Absorption | Surplus Component | Investment Income | Foreign Capital Assets | Total Capital Surplus |
|---|---|---|---|---|---|---|
| 1981 | 15.25 | 14.4 | 0.85 | 4.20 | 42.00 | 5.05 |
| 1982 | 14.61 | 16.4 | − 1.79 | 4.65 | 46.54 | 2.86 |
| 1983 | 11.55 | 17.2 | − 5.65 | 4.91 | 49.11 | − 0.74 |
| 1984 | 10.75 | 18.2 | − 7.45 | 4.84 | 48.44 | − 2.61 |
| 1985 | 10.09 | 18.8 | − 8.71 | 4.61 | 46.09 | − 4.10 |
| 1986 | 9.51 | 19.1 | − 9.59 | 4.24 | 42.40 | − 5.35 |
| 1987 | 9.13 | 19.2 | −10.07 | 3.75 | 37.58 | − 6.31 |
| 1988 | 8.73 | 19.3 | −10.57 | 3.19 | 31.89 | − 7.38 |
| 1989 | 8.55 | 19.4 | −10.85 | 2.52 | 25.25 | − 8.33 |
| 1990 | 8.48 | 19.5 | −11.02 | 1.77 | 17.75 | − 9.25 |
| 1991 | 8.35 | 19.7 | −11.35 | 0.94 | 9.42 | −10.41 |
| 1992 | 8.22 | 19.7 | −11.48 | 0.01 | 0.05 | −11.47 |
| 1993 | 8.07 | 20.2 | −12.13 | − 1.03 | − 10.27 | −13.16 |
| 1994 | 7.96 | 20.2 | −12.24 | − 2.22 | − 22.54 | −14.49 |
| 1995 | 7.86 | 20.2 | −12.34 | − 3.56 | − 35.58 | −15.90 |
| 1996 | 7.75 | 20.7 | −12.95 | − 4.99 | − 49.90 | −17.94 |
| 1997 | 7.64 | 20.6 | −12.96 | − 6.60 | − 66.04 | −19.56 |
| 1998 | 7.51 | 20.5 | −12.99 | − 8.36 | − 83.65 | −21.35 |
| 1999 | 7.42 | 20.9 | −13.48 | −10.28 | −102.86 | −23.76 |
| 2000 | 7.32 | 21.4 | −14.08 | −12.42 | −124.24 | −26.50 |

[a] For calculations, see notes to table 9.4.

position, on the other hand, shows a significant improvement with a diversion of investments from domestic to overseas, and the latter providing higher returns. One country that shows an improvement in these variables despite lower oil revenues is Saudi Arabia, whose current-account surpluses persist throughout the period. As for the United Arab Emirates, both capital surplus and foreign capital assets show increasing negative values. The aggregate values for the nations included in the scenario are summed up in table 9.19. It can be seen that even though total capital surplus shows a decline, total foreign capital assets increase, reaching $653.17 billion by the year 2000. Of course, the bulk of these assets are held by just two countries, Kuwait and Saudi Arabia, and the others generally find their overseas assets being reduced, the United Arab Emirates being the largest sufferer on this account.

In order to exhibit the sensitivity of these variables to price, I have computed their values based on oil price assumption $y$. A summary of these values is presented in table 9.20. Both oil revenues and foreign

TABLE 9.19

Scarcity Scenario, Price Assumption $x^a$, Middle East Capital Surplus OPEC Countries, 1981–2000 (billions of 1980 dollars)

| Year | Oil Revenue | Foreign Capital Assets | Total Capital Surplus |
|---|---|---|---|
| 1981 | 136.11 | 379.00 | 57.25 |
| 1982 | 121.21 | 430.53 | 24.76 |
| 1983 | 106.71 | 452.71 | 5.12 |
| 1984 | 113.37 | 457.36 | 5.00 |
| 1985 | 126.06 | 461.79 | 12.94 |
| 1986 | 126.79 | 473.37 | 14.34 |
| 1987 | 127.63 | 486.28 | 24.65 |
| 1988 | 128.70 | 508.43 | 16.44 |
| 1989 | 130.08 | 523.16 | 15.69 |
| 1990 | 131.85 | 537.16 | 18.56 |
| 1991 | 130.03 | 553.86 | 17.51 |
| 1992 | 127.55 | 569.60 | 13.61 |
| 1993 | 127.39 | 581.76 | 15.34 |
| 1994 | 126.40 | 595.04 | 12.30 |
| 1995 | 125.69 | 606.09 | 12.90 |
| 1996 | 125.02 | 617.70 | 10.59 |
| 1997 | 124.39 | 627.21 | 11.51 |
| 1998 | 123.69 | 637.55 | 9.54 |
| 1999 | 123.49 | 646.08 | 7.91 |
| 2000 | 123.94 | 653.17 | 6.76 |

[a] See table 9.1.

capital assets are much higher under this price assumption than under price scenario $x$ (table 9.19). After 1985, when a jump in oil prices is assumed, capital surplus also increases steadily. In other words, a single large jump in prices, given prudent economic management and policies, has significant benefits extending over the whole period.

The foregoing analysis provides an insight into OPEC's preference for "sudden" adjustments as opposed to an agreement on steady price increases. Any such agreement would, anyway, be virtually impossible to arrive at, since there are too many diverse interests to accommodate in a single agreement. Besides, producers and consumers in the world oil market differ about the time period to which the agreement would apply. For instance, major oil consumers are interested in stable prices over the short and medium term, which would permit them a transition to a postoil regime, whereas the motive of the oil exporters is to maximize the benefits from oil exports over a long time. This implies that oil exporters are concerned over the rates of depletion of their oil resources and the use of revenues and surpluses to modernize their economies to ensure steady and substantial incomes in the postoil era.

TABLE 9.20

Scarcity Scenario, Price Assumption $y^a$, Middle East Capital Surplus OPEC Countries, 1981–2000 (billions of 1980 dollars)

| Year | Oil Revenue | Foreign Capital Assets | Capital Surplus |
|---|---|---|---|
| 1981 | 136.11 | 379.00 | 57.21 |
| 1982 | 121.21 | 430.40 | 24.75 |
| 1983 | 106.71 | 452.67 | 5.18 |
| 1984 | 113.37 | 457.36 | 4.99 |
| 1985 | 156.01 | 461.79 | 42.89 |
| 1986 | 153.86 | 500.37 | 44.09 |
| 1987 | 151.79 | 540.03 | 54.17 |
| 1988 | 149.95 | 588.76 | 45.71 |
| 1989 | 148.64 | 629.89 | 44.92 |
| 1990 | 147.83 | 670.29 | 47.85 |
| 1991 | 145.86 | 713.33 | 49.29 |
| 1992 | 144.21 | 757.68 | 49.06 |
| 1993 | 142.83 | 801.81 | 51.71 |
| 1994 | 141.84 | 848.34 | 53.06 |
| 1995 | 140.86 | 896.08 | 57.07 |
| 1996 | 140.12 | 947.41 | 58.66 |
| 1997 | 139.39 | 1,000.20 | 63.80 |
| 1998 | 138.68 | 1,057.61 | 66.53 |
| 1999 | 138.42 | 1,117.48 | 69.97 |
| 2000 | 138.94 | 1,180.43 | 75.47 |

[a] See table 9.1.

The assumption of lower growth in the nonoil gross domestic product of the OPEC core nations under the scarcity scenario has an eminently rational basis. First, the ill effects caused by overcommitment of expenditure in anticipation of higher oil revenues, which we witnessed in most OPEC nations in the 1970s, has produced some lasting lessons. Second, in the decade since 1973, OPEC members have increasingly concluded that economic development takes much more than a mere infusion of capital. They realize that much of the oil revenues earned in the post-1973 period has been imprudently spent, with meager long-term benefits. Third, the post-1981 glut in the oil market has induced a conservatism in oil production and exports, which will limit growth in oil exports and revenues for many years to come. For these reasons, and the likely higher benefits to the core OPEC nations as evidenced by the values in table 9.20, I feel that this particular scenario will hold greater appeal for these countries than the other scenarios examined. In the coming years, therefore, they are likely to try to create conditions that would bring about such a scenario.

In actual fact, the lower oil output and export levels in the scarcity scenario would actually result in a higher rate of growth of oil prices and, hence, greater revenues and surpluses. It seems inevitable that OPEC will gradually reduce production to increase its revenues, to lengthen the life of its oil reserves, and to slow down its rate of development. It can be concluded, therefore, that the future may lie closer to the scarcity scenario than the blinkers-on scenario.

## CHAPTER 10
# CONCLUSIONS

Most of the analysis in earlier chapters appears to justify the hypothesis of strong two-way linkages among OPEC, the industrial nations, the Third World, and the international economic order. In a world of growing internationalism, the understanding of linkages among groups of nations is particularly important in evaluating future global developments. Often in the past, these linkages have not been fully understood by researchers and policymakers, who have preferred to concentrate on narrow issues of direct national concern. The limitation of such an approach has been fully established by the major (often violent) fluctuations in the international energy market and in global economic relations. For instance, few people believed in the early 1970s that oil prices would quadruple in the short period between 1973 and 1974, and perhaps an equally small number of people could have predicted in 1979–80 that oil prices would actually decline in nominal terms in early 1983. Further surprises may be in store for tomorrow's civilization if some of the lessons from recent global developments are not understood and accepted by today's leaders and decision makers. Researchers, therefore, have a responsibility when evaluating the political economy of global energy to supersede traditional models and methodologies.

The high growth in the demand for energy witnessed during the 1950s and 1960s is over. The growth rate for energy demand in the 1970s set a trend that is not likely to alter significantly in the future. Sluggish economic growth in the industrial nations is likely to be a rule rather than an exception during the rest of the 1980s and perhaps a good part of the 1990s, but it is unlikely that demand will, in the long run, fall below the 2 percent rate of growth projected by the U.S. Department of Energy (1981, 2–6). It may happen that during this period the industrial nations will be able to restructure their economic systems, developing a technological base which will give them greater sophistication and superiority and a spurt in economic growth. But it is unlikely that nations of the Third World, some of which have already bridged the industrial and technological gap, would stay far behind. Consequently, the comparative advantage of technology and abundant capital (both physical and human) and declining real prices of energy, which worked to the benefit of the industrial nations in the 1950s and 1960s, will not hold in the future. Oil prices in particular are undoubtedly going to move upward in the long run, despite the decline of 1983 and its underlying causes. Lower economic growth in the industrial nations, quite apart from reducing the growth in demand for oil, would have a profound impact on the economies of the Third World nations, particularly the poorest, which have no smooth recourse to borrowing from the international banks. Lower economic growth in the OECD nations means lower levels of official development assistance, as is being witnessed currently. Hence, the vigorous foreign aid programs that marked the 1950s and 1960s are also perhaps over. The poorest nations in the Third World are, therefore, likely to turn more and more to the richer nations of the Third World, particularly OPEC nations which, despite depressed oil market conditions, will continue to generate substantial current-account surpluses.

The emergence of OPEC as the prime mover in world oil supplies and trade is a factor which will continue to cast its shadow over the world's energy future and economic health. This fact is of particular importance, because in the past many in positions of power adopted an attitude of indifference, even arrogance, to the leaders of this group of nations. But they reached their position of strength on the basis of history, and it is not likely to be abandoned easily. No doubt, OPEC has weaknesses, of which no greater evidence is necessary than the internal squabbles and disarray of 1981–82. Also, OPEC's pricing decisions of 1979–80 appear to have been an expression of immaturity and haste, which are likely to prove harmful to its own interests in due course.[1] But as my analysis has put forward, OPEC's strength lies in the excess capacity of Saudi Arabia and its willingness to vary output to influence prices and quantities. And as long as OPEC remains an effective force in the world oil market, the actions of Saudi Arabia and, to a smaller degree, the other capital-surplus nations of the Middle East will be of great consequence to global economic developments. Political instability in the region can bring about internal strains in this body, but it is unlikely to reach a dimension that would break up the "cartel." In the worst possible scenario, if there is an irreconcilable split among the members

---

1. Referring to the oil price rises of 1979–80, Sheikh Yamani is reported to have said, "We were almost intoxicated by the fact that we could raise our prices the way we wanted." "Hard Lessons for Exporters," *Petroleum Economist* 49(3), 83.

of this body, there is every reason to expect that it may be replaced by a smaller and more homogeneous grouping (Eckbo 1976). Such a group would include Saudi Arabia, Kuwait, the United Arab Emirates, Qatar, and possibly Iraq. These five nations could export up to 15 million barrels of oil a day in the near future. It is unlikely that the other eight members could form an effective force with any significant control over the market. However, it is highly unlikely that the situation would reach such a crisis, even though a great deal of shadowboxing may continue, with Iran, Nigeria, or Libya defying temporary production quotas.

Meanwhile, the poorest nations of the world would continue to face grave difficulties. (1) Their oil import burdens, even with depressed oil market conditions, would continue to be unbearable. (2) Their overseas debts are much too large in relation to export earning capabilities. (3) Weak commodity markets, for which historic reasons are as important as the recessionary conditions prevailing in the industrial nations, will not permit them to increase their export earnings to a level where the first two problems can be solved. There are, therefore, strong links among the conditions of the oil market, the economic health of the industrial nations, and the economic survivability of Third World nations. These links are related not only to current conditions but to a series of lagged effects that go back in time. Consequently, even if effective remedies were to be instituted today, the effects of these will not be seen for years to come. The debt problem of the Third World lies not only with the borrowing nations, which have done little to readjust their plans and economic policies, but also with the private banking community. OPEC's unprecedented surpluses of the mid-1970s found a convenient outlet in the private banking system. After the immediate borrowing needs of the industrial nations had been met, the private banks, flush with easy OPEC money, looked to other outlets, and these were conveniently available in other Third World countries, particularly in Latin America. These darlings of the private banking system of yesterday have become the *enfants terrible* of today. The only long-term solution, namely a hard and painful adjustment, was not undertaken in the mid-1970s when it would have done the most good. Jeffrey Garten's (1983) recipe for solving the Third World debt problem includes such an adjustment plus the close cooperation of organizations such as the International Monetary Fund and the World Bank with the private banking system to ensure the implementation of readjustment plans and policies. The Third World's tight financial crunch will inevitably lead to defaults on existing debts, lower economic growth, and increasing political-social turmoil. The problem is not likely to reach crisis proportions in the aggregate, but individual countries will continue to have serious difficulties in paying their international debts.[2] Continuing problems of this kind are at the root of the ongoing North-South dialogue and the demand for a new international economic order.

The United Nations advocates income transfers from North to South. Besides the program advocated by the Brandt Commission, a program for stabilizing commodity prices and improving the terms of trade for commodity exporters has been put forward. It is unlikely that any international agreement can be arrived at for artificially propping up commodity prices and reversing the decline of the past few decades, but there is considerable merit in a price stabilization program based on an internationally financed buffer stock. Simulations by Jere Behrman (1978) reveal that the cost of such a program, with a 15 percent band width for price stabilization, would have been only $1.892 billion for the period 1962–73. Revenue stabilization would have enormous benefits to Third World countries dependent on commodity exports, because they would be able to plan more effectively and allocate resources more efficiently for economic development. It should not be difficult for OPEC, as the exporter of the most valuable commodity today and with experience of the ill effects of international commodity pricing, to bankroll such a program.

Currently, the major cause of imbalances in the international monetary system, namely the surplus generated by OPEC, is itself going through a period of rapid change. This is being brought about not only by decline in revenues but by a reappraisal of development performance by the OPEC nations themselves, which are aware of the waste and misallocation of resources from the oil revenues of the mid-1970s. Consequently, even though per capita income has gone up dramatically in some OPEC nations, particularly in the Persian Gulf region, other indicators of development have not changed substantially. For instance, education, technical skills, and services have not kept pace with increases in per capita income. Consequently, most of these nations remain dependent on other countries for capital goods as well as consumer goods.

Another factor influencing the downward adjustment of oil production and exports, particularly in the OPEC core nations, is the realization of the finite nature of oil resources. This factor is a cause of some dissension among groupings of members within OPEC. The more populous high absorbers adopt hawkish postures, wanting to step up oil output and revenues, whereas the low absorbers adopt a more moderate approach. Despite these differences, OPEC has continued to remain

---

2. This is essentially the message in a study briefly reported in the *Sloan Management Review* and *Best of Business*. See Zenoff and Howard.

remarkably unified on matters of major policy. In fact, OPEC managed to live reasonably well with rigid production quotas, which generally herald the demise of cartels of the classical type. This only establishes the thesis presented earlier—that OPEC is not truly a cartel but conforms more closely to a dominant-firm price-leadership model. It is difficult to predict whether this grouping will continue intact, but if the experience of the last ten years is any indication, OPEC will accommodate future disturbances.

The implication of a lower growth strategy and lower levels of export by OPEC is that surpluses will be much lower in the coming years, but the scenarios based on these lower growth strategies indicate that OPEC's surplus will not suffer in the long run if a modest strategy is maintained in nonoil sector investments. There will undoubtedly be imbalances among the member countries, and this is likely to result in a greater amount of aid and lending by the capital-surplus nations of OPEC to the members that incur deficits. Critical elements in future developments are (1) the extent of non-OPEC oil production, which will be constrained by lack of financial and other resources in the Third World, and (2) the effects of energy conservation programs, which have, at least for the present, reached close to their anticipated potential.

OPEC assets in the industrial nations do not in real terms fetch high returns, even though they do provide a high degree of security. OPEC and the poorer nations of the Third World may, therefore, contract a marriage of convenience, resulting in larger assistance, concessional lending, and other flows to these Third World nations. Also, the growth of Arab banking may lead to their channelling funds and diversifying portfolios into at least those nations of the Third World with robust industrial structures and institutions. These developments, however, depend on the maturity and statesmanlike behavior of core OPEC nations, not always evident in the past. Irrespective of the financial commitment, there is little doubt about OPEC's political commitment to the Third World. OPEC is, therefore, likely to continue to support a new international economic order. Also, parts of the agenda that other Third World countries put forward fall directly within the range of OPEC's interests. These include the demand for reform of the international monetary system and acceptance of the principle that commodities should fetch higher prices in the long run in markets of the North.

A more universal and mature reason for OPEC to provide other nations of the Third World with developmental finance is that it is good politics. First, OPEC will need the support of the rest of the Third World in future oil price increases. Second, the international banking system is unlikely to be able to handle petrodollar flows of the size that followed the 1973–74 and 1979–80 price increases. OPEC will have to find channels and develop a much larger portfolio of investments if its surpluses are to be efficiently recycled and if its capital assets are to be welcome in the recipient nations. Thus far Kuwait, among all the capital-surplus nations, has been businesslike in its investments and has not hesitated in direct-equity participation overseas (although most of its investments have gone to industrialized nations, obviously in view of perceived lower risks). An area in which OPEC's active role is desirable is in joint projects in the Third World, particularly in countries with a record of stability and efficiency in organized production.

The most heartening development in the Third World in recent decades has been in the increase in manufactured goods for export. The long-run solution for commodity exporters is diversification by investments in other industries. Unfortunately, only very few Third World nations have been able to achieve success in this area, although the all-round record has been noteworthy. The major hindrances are the growing protectionism of the industrial nations and the lack of finance, technology, infrastructure, and skills in the poorest of the commodity producers. An increase in development assistance for the poorest nations, as vehemently argued by the Brandt Commission, is in the interests of global stability. The importance of the global effects of enlightened aid policies and the consequences of economic and social upheaval in the Third World have to be understood by leaders of the richer nations, including OPEC.

But in no area of global development is the argument of a public good more applicable than in the case of energy production. Given the relationship between availability of energy resources and the costs of energy production, it is in the overall interests of a world dependent on international energy trade to tap the least-cost sources of energy wherever they exist. The growing scarcity of global energy and likely reductions in supply can only be countered by a massive effort to produce more energy from newer sources. Oil-importing nations of the West would find in this approach a means for steadying oil prices, and even OPEC would benefit in the long run from lower pressure on their exports, because it would reduce the political and military threats that could arise if supplies fall much below demand. The largest beneficiaries would, of course, be the oil-importing developing countries, some of which have significant energy resources but inadequate finance for their exploitation. The idea of a large energy-funding organization has been proposed in various forms, but it would not be desirable to develop a single bureaucratic juggernaut in this field. Instead, existing institutions and channels could be used for increased funding and local expertise could be developed to implement and manage these programs.

These policies and strategies and their likely global implications can, however, be appreciated only if the future of energy demand and supply can be assessed—with particular reference to oil, which is bound to continue dominating energy use into the next century. The special feature of the oil problem lies in the ability and willingness of OPEC to continue exporting at levels that would fill the residual gap between oil demand and supply for the noncommunist world. The implications of the scenarios I constructed to assess OPEC's actions in the future provide evidence of tightening oil markets in the nineties. (The figures in these scenarios are not predictions but merely orders of magnitudes of variables, whose interlinkages are often ignored in assessments of oil market developments.) In developing these scenarios, I assumed a reasonable increase in real oil prices and postulated growth rates for nonoil gross domestic product in the capital-surplus members of OPEC on the basis of their economic structures. The growth of their surpluses and overseas capital assets were then charted in consonance with domestic development policies and rates of economic growth. Within this framework, the blinkers-on scenario reveals a rate of buildup of annual surpluses far beyond the recycling capacity of the international banking system and a rate of increase of overseas assets, which could lead to serious political problems in the countries in which these assets are held. I believe that the rates of oil production characterizing the blinkers-on scenario will not be attained by an overt policy of lower production aimed at slower economic growth in the capital-surplus nations and slower growth of capital surpluses.

In the scarcity scenario, rates of growth of nonoil gross domestic product are correspondingly lower and more acceptable, and oil output levels are more likely to happen than those in the blinkers-on scenario. First, the lower rates of growth of nonoil gross domestic product are likely to minimize social and economic problems in these nations. Second, there would be hardly any reduction in the growth of overseas assets, in contrast to the blinkers-on scenario. Third, the conversion of oil resources into financial assets, which has in any case been yielding poor financial returns, would be slower, thereby conserving oil reserves for a longer period and giving them a longer time for a postoil economic transition. It needs to be emphasized that the OPEC capital-surplus nations, dominated by Saudi Arabia, are the core of OPEC in terms of output and exports. These are also nations beset with actual or likely political instability. The reappraisal of their experience since 1973 underway in these nations, is bound to lead to development plans more cautious than the grandiose experiments of the seventies. The problem of migrant labor and the dominance of foreigners in their populations is going to be a particularly important influence in constraining future growth.

Extending the scarcity scenario, we arrive at total OPEC output of 20.3 million barrels of oil a day in 1985, 19.8 in 1990, and 16.1 in 2000. These figures are far below the projections shown in table 3.9 of 27.1, 27.2, and 26.2 mbd respectively in the same years. The implication is that, if supply shown in table 3.9 is required to meet demand, then there may actually be a shortfall or excess demand for oil of 6.8 million barrels a day in 1985, 7.4 in 1990 and 10.1 in 2000. Such an outlook would cause severe crises in the rest of the world but would actually be beneficial to OPEC for its effects on oil prices. Consequently, the scarcity scenario could conceivably be revised downward for oil output without any loss in revenue, economic growth, surpluses, or buildup in capital assets for the capital-surplus nations. However, their spending spree would certainly be moderated in the next two decades, stemming from an unfavorable assessment of past policies. Such a lowering of domestic growth along with high oil output would only result in abnormally large surpluses, which would choke the international banking system and, in effect, have no outlets for productive investments. The deficits of the oil-importing developing countries are of a chronic nature and will not disappear without a restructuring of their economies, and their demands for financing will continue to be accompanied by a commensurate inability to repay. On the external front, therefore, OPEC may find it beneficial to either diversify its investments radically or to cool down the rate of flow of petrodollars to other nations. The lesson from these prospects appears very clear: oil output in OPEC will most definitely dip below 20 million barrels a day in the 1990s.

It would be fitting to end this book by putting forward the elements of a global strategy that would avoid this impending crisis. I believe that the following agenda, if accepted and implemented, would have positive global effects in mitigating the effects of these likely developments:

1. A vigorous program of energy conservation is important all over the world. National policies must be based on the public-good concept, with systems of taxes and subsidies to promote conservation efforts toward efficiency in a social sense. This program needs to be extended to the developing countries as well, where sectors such as transport and industry are highly wasteful of energy resulting from subsidized energy prices. There are large inefficiencies in the use of traditional fuels, as well, particularly where abundant supplies are available without monetary cost. Such areas are in danger of losing forest and other biomass resources, as has

happened in Nepal and Sahelian Africa. Indeed, as most recent works on the subject indicate, conservation is the most important source of energy "supply."
2. An active program for enhanced financing of energy developments in the Third World needs to be instituted, with greater funding from the industrial nations and OPEC. The Brandt Commission proposal for a tax on international energy trade has considerable merit, since it could actually reduce energy consumption by effectively raising the price of imported energy and would provide finances for developing energy resources in other countries of the Third World.
3. Critical to the stability of the international economic order is steady economic growth in the Third World. Calls for a new international economic order need to be taken seriously by the countries of the North and followed up with rhetoric-free and constructive negotiations by the South. Apart from removing barriers to imports from the Third World, the industrial nations must be willing to establish and fund a buffer-stock price-stabilization program for commodities. OPEC, as the most successful group of commodity exporters, must take a more active role in financing such a program. The nations of the South should give up demands for higher prices for commodities, since this would neither be acceptable to nations of the North nor be administratively feasible to implement.
4. The surpluses of OPEC need to be channeled through alternative channels on a large scale, since the international banking system is not likely to have the capacity to handle future surpluses. Arab banking, which shows promise, may be able to take on a more aggressive role in financing projects in the Third World. A shift in geographical distribution of future portfolios would also minimize the risks of exchange-rate fluctuations and political problems, such as the freezing of Iranian assets by the U.S. government. The size of future asset accumulation makes this feature particularly important.
5. OPEC must play a larger global role consistent with its power in the world energy market and its financial wealth. Rather than be viewed as a source of insecurity and instability, it could emerge as a stabilizing force in the world economy. Its world presence would be based on an enlightened and broad-minded aid policy, broken from the confines of Islamic bias.

These ideas have been discussed and debated in various forums, but their likely consequences have not been fully accepted because the absence of such actions have not been evaluated within a global framework. It is hoped that the analyses of the foregoing pages make clear the urgency of these measures and that the influential institutions of the world consider them seriously.

# SELECT BIBLIOGRAPHY

Adams, F. G., and J. R. Behrman. 1976. *Econometric Models of World Agricultural Commodity Markets: Cocoa, Coffee, Tea, Wool, Cotton, Sugar, Wheat, Rice.* Cambridge, Mass.: Ballinger.

Adelman, M. A. 1972. *The World Petroleum Market.* Baltimore: Johns Hopkins University Press.

Adelman, M. A. 1980. "The Clumsy Cartel." *Energy Journal* 4 (3).

Adelman, M. A. 1981. "Coping with Supply Insecurity." Presidential Address, Third Annual Meeting of the International Association of Energy Economists.

Agmon, Tamir, Donald R. Lessard, and James L. Paddock. 1981. "Financing Petroleum Development in Developing Countries." MIT-E1 81–059 WP. Cambridge, Mass.: MIT Energy Laboratory.

Agnew, M., L. Schrattenholzer, and A. Voss. 1979. "A Model for Energy Supply Systems Alternatives and their General Environmental Impact." WP-79-6. Laxenburg, Austria: International Institute for Applied Systems Analysis.

Akins, James E. 1973. "The Oil Crisis: This Time the Wolf is Here." *Foreign Affairs*, April.

Akins, J. E. 1977. "World Energy Supply: Cooperation with OPEC on a New War for Resources." Paper presented at the Third International Symposium on Petroleum Economics.

Alegria, F., and others. 1983. "Influence of Energy Conservation and New Technologies on Spain's Energy Balance for the Year 2000." Paper presented at the Twelfth Congress of the World Energy Conference.

Al-Janaby, Adnan A. 1980. "Equilibrium of External Balances Between Oil Producing Countries and industrialized Countries." *OPEC Review* 3(4) and 4(1).

Amacher, Ryan C., Gottfried Haberler, and Thomas D. Willett, eds. 1979. *Challenges to a Liberal International Economic Order.* Washington, D. C.: American Enterprise Institute.

Aman, Fernando, ed. 1981. *Energy Demand and Efficient Use: Proceedings of the Fourth International School on Energetics.* New York: Plenum.

Amin Galal, A. 1974. *The Modernization of Poverty.* Leiden, Netherlands: E. J. Brill.

## 180 / Bibliography

Aperjis, D. 1982. *Oil in the 1980's: OPEC Oil Policy and Economic Development.* Cambridge, Mass.: Ballinger.

Arismunandar, A. 1975. "Indonesian Energy Demand in the Year 2000." Paper presented at the Symposium on Energy, Resources and Environment.

Attiga, Ali A. 1983. "Energy Development in the Arab World: Present Situation and Future Prospects." Paper presented at the Twelfth Congress of the World Energy Conference.

Attiga, Ali A. 1984. "Price of Oil and the Energy Market." Paper presented at the Fifth International Conference of the International Association of Energy Economists.

Australian Department of National Development and Energy. 1982. *Energy Forecasts for the 1980s.* Canberra: Australian Government Publishing Service.

Balassa, Bela. 1981. *The Newly Industrializing Countries in the World Economy.* New York: Pergamon.

Barnett, Harold J., and Chandler Morse. 1963. *Scarcity and Growth: The Economics of Natural Resources Availability.* Baltimore: Johns Hopkins University Press for Resources for the Future.

Bartlett, Geoffrey. 1981. *Rising Oil Prices and World Economic Output.* London: Economist Intelligence Unit.

Basile, Paul S., ed. 1977. "Energy Supply-Demand Integrations to the Year 2000: Global and National Studies." Third technical report, workshop on Alternative Energy Strategies. Cambridge, Mass.: Massachusetts Institute of Technology.

Behling, D. J., R. Dullien, and E. A. Hudson. 1976. "The Relationship of Energy Growth to Economic Growth under Alternative Energy Policies." BNL 50500. Upton, N. Y.: Brookhaven National Laboratory.

Behrman, Jere R. 1978. *Development, the International Economic Order, and Commodity Agreements.* Reading, Mass.: Addison-Wesley.

Berndt, E. R., and L. R. Christensen. 1973. "The Internal Structure of Functional Relationships: Separability, Substitution and Aggregation." *Review of Economic Studies* 40, July.

Berndt, E. R., and D. O. Wood. 1976. "Technology, Prices, and the Derived Demand for Energy." *Review of Economics and Statistics* 58 (1).

Bhagwati, Jagdish N., ed. 1977. *"The New International Economic Order: The North-South Debate.* Cambridge, Mass.: MIT Press.

Bijan, Mossavar-Rahmani, and Fereidun Fesharaki. 1983. "OPEC and the World Oil Outlook: Rebound of the Exporters." Special Report 140. London: Economist Intelligence Unit.

Blair, John M. 1976. *Control of Oil.* New York: Pantheon.

Bohm, Robert A., and others, eds. 1981. *World Energy Production and Productivity: Proceedings of the International Energy Symposium, 1980.* Cambridge, Mass.: Ballinger.

Brookhaven National Laboratory. 1978. "Energy Needs, Uses and Resources in Developing Countries." Upton, N.Y.

Cazalet, E. Z. 1977. *Generalized Equilibrium Modelling: The Methodology of the SRI-Gulf. Energy Model.* Palo Alto, Calif.: Decision Focus.

Chenery, Hollis B., and Donald B. Keesing. 1979. "The Changing Composition of Developing Country Exports." Working Paper 314. Washington, D.C.: World Bank.

Cherniavsky, E. A. 1974. "Brookhaven Energy System Optimization Model." BNL 19569. Upton, N.Y.: Brookhaven National Laboratory.

Choucri, Nazli. 1981. *International Energy Futures: Petroleum Prices, Power and Payments.* Cambridge, Mass.: MIT Press.

Connolly, T. J., George B. Dantzig, and S. C. Parikh. 1979. "The Stanford PILOT Energy/Economic Model." In *Advances in the Economics of Energy and Resources*, vol. 3, edited by Robert S. Pindyck. Greenwich, Conn.: JAI Press.

Cook, P. Lesley, and A. J. Surrey. 1977. *Energy Policy: Strategies for Uncertainty.* London.: Martin Robertson.

Cremer, J., and H. Weitzman. 1976. "OPEC and the Monopoly Price of World Oil." *European Economic Review* 8.

Daig, R. H., and W. K. Tabb. "A Dynamic Micro-economic Model of the US Coal Mining Industry." SSRI Research Paper. Madison: University of Wisconsin.

Daly, George, James M. Griffin, and Henry B. Steele. 1983. "The Future of OPEC: Price Level and Cartel Stability." *Energy Journal* 4 (1).

Dantzig, G. B., and S. C. Parikh. 1975. "On a PILOT Linear Programing Model for Assessing Physical Impact on the Economy of a Changing Energy Picture." In *Energy: Mathematics and Models, Proceedings of SIMS Conference on Energy*, edited by Fred S. Roberts.

Darmstadter, Joel. 1971. *Energy in the World Economy.* Baltimore: Johns Hopkins University Press.

Dasgupta, P., and G. M. Heal. 1974. "The Optimal Depletion of Exhaustible Resources." *Review of Economic Studies Symposium.*

Dunkerley, Joy, ed. 1980. *International Energy Strategies: Proceedings of the 1979 International Association of Energy Economists.* Cambridge, Mass.: Oelgeschlager, Gunn, and Hain.

Ebinger, Charles K. 1981. *Pakistan: Energy Planning in a Strategic Vortex.* Bloomington: Indiana University Press.

Eckbo, Paul Lee. 1976. *The Future of World Oil.* Cambridge, Mass.: Ballinger.

Edmonds, J., and others. 1981. *Determinants of Global Energy Supply to the Year 2050.* Oak Ridge, Tenn.: Oak Ridge Associated Universities.

Edmonds, J., and J. Reilly. 1983*a*. "A Long Term Global Energy Economic Model of Carbon Dioxide Release from Fossil Fuel Use." *Energy Economics* 5 (2).

Edmonds, J., and J. Reilly. 1983*b*. "Global Energy Production and Usage in the Year 2050." *Energy* 8(6).

"Energy Economists Unite." 1983. *Energy Detente* 4(15).

Energy Modelling Forum. 1977. "Energy and the Economy." EMF Report 1. Stanford: Stanford University.

Energy Modelling Forum. 1982. "World Oil." EMF Study 6. Stanford: Stanford University.

Ericsson, N. R., and P. Morgan. 1978. "The Economic Feasibility of Shale Oil: An Activity Analysis." *Bell Journal of Economics* 9(2).
Exxon Corporation. 1979–1981. "World Energy Outlook." New York.
Ezzati, Ali. 1978. *World Energy Markets and OPEC Stability*. Lexington, Mass.: D. C. Heath.
Fesharaki, Fereidun. 1976. *Development of Iranian Oil Industry*. New York: Praeger.
Fesharaki, Fereidun. 1981. "World Oil Availability: The Role of OPEC Policies." *Annual Review of Energy* 6.
Fesharaki, Fereidun, and David T. Isaak. 1981 *OPEC Downstream Processing—A New Phase of the World Oil Market*. Honolulu: Resource Systems Institute.
Freeman, Christopher, and Marie Jahoda, eds. 1978. *World Futures: The Great Debate*. Sussex, England: Science Policy Research Unit.
Freeman, David S. 1974. *A Time to Choose*. Cambridge, Mass.: Ballinger.
Garten, Jeffrey E. 1983. "LDC Debt: Toward More Innovative Approaches." *Economic Impact: A Quarterly Review of World Economics* no. 43.
Griffin, James M. 1979. *Energy Conservation in the OECD: 1980 to 2000*. Cambridge, Mass.: Ballinger.
Griffin, J. M., and P. R. Gregory. 1976. "An Intercountry Translog Model of Energy Substitution Responses." *American Economic Review* 66 (5).
Hafele, Wolf. 1981 *Energy in a Finite World: Paths to a Sustainable Future*. Cambridge, Mass.: Ballinger.
Hallwood, Paul, and Stuart Sinclair. 1981. "An Interpretation of the Economic Relationships Between OPEC and Non-Oil LDC's During the 1970's." *OPEC Review* 5 (3).
Haq, Mahboob Ul. 1980. "Negotiating the Future." *Foreign Affairs* 59 (2).
Hardy, Randall W. 1978. *China's Oil Future: A Case of Modest Expectations*. Boulder, Colo.: Westview Press.
Henderson, P. D. 1975. *India: The Energy Sector*. Oxford: Oxford University Press.
Hoffman, K. C., and D. W. Jorgenson. 1977. "Economic and Technological Models for Evaluation of Energy Policy." *Bell Journal of Economics* 8(2).
Hoffman, Thomas, and Brian Johnson. 1981. *The World Energy Triangle: A Strategy for Cooperation*. Cambridge, Mass.: Ballinger.
Hogan, W. W. 1983. "Oil Stockpiling: Help Thy Neighbour." *Energy Journal* 4 (3).
Hogan, W. W., and A. S. Manne. 1977. "Energy-Economy Interactions: The Fable of the Elephant and the Rabbit." In *Modelling Energy-Economy Interactions: Five Approaches*, edited by C. J. Hitch. Washington, D.C.: Resources for the Future.
Hollander, Jack, and others, eds. 1976–84. *Annual Review of Energy*. Vols 1–8. Palo Alto, Calif.: Annual Reviews.
Hotelling, Harold, 1931. "The Economics of Exhaustible Resources." *Journal of Political Economy* 39(2).
Hudson, E. A., and D. W. Jorgenson, 1974. "U.S. Energy Policy and

Economic Growth, 1975–2000." *Bell Journal of Economics and Management Science* 5(2).

Humphrey, David B., and J. R. Moroney. 1975. "Substitution Among Capital, Labour, and Natural Resource Products in American Manufacturing." *Journal of Political Economy* 83, February.

Independent Commission on International Development Issues. 1980. *North-South: A Programme for Survival.* Cambridge, Mass.: MIT Press.

International Energy Agency. 1979. Workshop on Energy Data of Developing Countries, 1978. Vol. 1 *Summary of Discussions and Technical Papers.* Vol. 2 *Basic Energy Statistics and Energy Balances of Developing Countries, 1967–1977.* Paris: OECD.

International Energy Agency. 1981. *Energy Conservation: Role of Demand Management in 1980s.* Paris: OECD.

International Energy Agency. 1982. *World Energy Outlook.* Paris: OECD.

International Energy Agency. 1983. *Energy Statistics of OECD Countries.* Paris: OECD.

International Institute for Applied Systems Analysis. 1981. *Energy in a Finite World: Executive Summary.* Report by the Energy Systems Programme Group of the IIASA. Laxenburg, Austria.

Ion, D. C., 1980. *Availability of World Energy Resources.* 2d ed. London: Graham and Trotman.

Jacobson, Harold K., Dusan Sidjanski, Jeffrey Rodamar, and Alice Hougassian-Rudovich. 1983. "Revolutionaries or Bargainers? Negotiators for a New International Economic Order." *World Politics*, April.

Jacoby, Henry D., and James L. Paddock. 1981. "World Oil Prices and Economic Growth in the 1980s." MIT-EL 81-060 WP. Cambridge, Mass.: MIT Energy Laboratory.

Jacoby, N. H. 1974. *Multinational Oil.* New York: Macmillan.

Jahangir, Amuzegar. 1983. *Oil Exporters: Economic Development in an Interdependent World.* Washington, D. C.: International Monetary Fund.

Jorgenson, D. W., and K. G. Hoffman. 1977. "Economic and Technological Models for Evaluation of Energy Policy." *Bell Journal of Economics* 8(2).

Jorgenson, D. W., and B. D. Wright. 1975. "The Impact of Alternative Policies to Reduce Oil Imports." *Data Resources Review* 4(6).

Kennedy, Michael. 1974. "An Economic Model of the World Oil Market." *Bell Journal of Economics and Management Science*, Autumn.

Khazzoom, J. D. 1976. "Proceedings of the Workshop on Modelling the Inter-Relationships Between the Energy Sector and the General Economy." Special Report 45. Palo Alto, Calif.: Electric Power Research Institute.

Kononov, Y., and A. Por. 1979. "The Economic Impact Model." RR-79. Laxenburg, Austria: International Institute for Applied Systems Analysis.

Krapels, E. 1980. *Oil Crisis Management: Strategic Stockpiling for International Security.* Baltimore: Johns Hopkins University Press.

Krauss, Melvyn B. 1980. "Brandt Report Is Irrelevant to Third World." *Wall Street Journal*, August 18.

Kursunoglu, Behram N., and others, eds. 1983. *Energy for Developed and*

*Developing Countries: Proceedings of the International Scientific Forum on Energy for Developed and Developing Countries, 1979.* Lexington, Mass.: Lexington Books.

Labys, Walter C. 1982. "A Critical Review of International Energy Modelling Methodologies." Working Paper 44. Morgantown, W.Va.: Department of Mineral Energy and Resource Economics, West Virginia University.

Lamb, Richard. 1982. "Making the Energy Transition." *Energy Policy* 10 (1).

Lambertini, Adrian, and others. 1976. "Energy and Petroleum in Non-OPEC Developing Countries, 1974–1980." Staff Working Paper 229. Washington, D.C.: World Bank.

Landsberg, Hans H., ed. 1979. *Energy: The Next Twenty Years.* Cambridge, Mass.: Ballinger.

Landsberg, Hans H., and Joseph H. Dukert. 1981. *Energy Costs: Uneven, Unfair, Unavoidable.* Baltimore: Johns Hopkins University Press.

Lapillonne, Bruno. 1978. *Medeez: A Model for Long Term Energy Demand Evaluation.* Laxenburg, Austria: International Institute for Applied Systems Analysis.

Lonnroth, Mans, and others. 1977. *Energy in Transition: Report on Energy Policy and Future Options.* Translated by Rudy Feichtner. Stockholm: Secretariat for Future Studies.

Lukachinski, Joan, P. J. Groncki, and R. G. Tessmer, Jr. "An Integrated Methodology for Assessing Energy Economy Interactions." BNL 26452. Upton, N.Y.: Brookhaven National Laboratory.

Mabro, Robert, ed. 1980. *World Energy Issues and Policies: Proceedings of the First Oxford Energy Seminar.* Oxford: Oxford University Press.

Maidique, Modesto A. 1979. "Solar America." In *Energy Future*, edited by Robert Stobaugh and Daniel Yergin. New York: Random House.

Makrakis, Michael S., ed. 1974. "Energy: Demand, Conservation, and Institutional Problems: Conference Proceedings." Cambridge, Mass.: Massachusetts Institute of Technology.

Mallakh, Ragaei El, ed. 1982. *OPEC: Twenty Years and Beyond.* Boulder, Colo.: Westview Press.

Mangone, Gerard J., ed. 1976. *Energy Policies of the World: Canada, China, Arab States of the Persian Gulf, Venezuela, Iran.* New York: Elsevier.

Mangone, Gerard J., ed 1977. *Energy Policies of the World: Indonesia, the North Sea Countries, the Soviet Union.* New York: Elsevier.

Mangone, Gerard J., ed. 1979. *Energy Policies of the World: India, Japan.* New York: Elsevier.

Manne, A. 1977. "ETA Macro: A Model of Energy-Economy Interactions." Paper presented at ORSA/TIMS meeting.

Marcuse, W., L. Bodin, E. A. Cherniavsky, and Y. Sanborn. 1977. "A Dynamic Time Dependent Model for the Analysis of Alternative Energy Policies." BNL 19406. Upton, N.Y.: Brookhaven National Laboratory.

Mathematical Sciences Northwest. 1977. "North-West Energy Policy Project: Energy Demand Forecasting Model, Technical Appendix." Report NEPP 11 (PB-276921). Washington, D.C.: Mathematical Sciences Northwest.

Meier, Peter, and V. Mubayi. 1981, "Modelling Energy-Economic Interactions in Developing Countries: A Linear Programming Approach." BNL 29747. Upton, N.Y.: Brookhaven National Laboratory.
Miernyk, W. H. F. Guarratam, and C. F. Socher. 1978. *Regional Impacts of Rising Energy Prices*. Cambridge, Mass.: Ballinger.
Mikdashi, Zuhayr. 1972. *The Community of Oil Exporting Countries: A Study in Governmental Cooperation*. Ithaca, N.Y.: Cornell University Press.
Moran, Theodore H. 1978. *Oil Prices and the Future of OPEC: Political Economy of Tension and Stability in the Organisation of Petroleum Exporting Countries*. Washington D.C.: Resources for the Future.
Moroney, John R., and Alden L. Toevs. 1979. "Input Prices, Substitution, and Product Inflation." In *Advances in the Economics of Energy and Resources*, vol. 1, edited by Robert S. Pindyck. Greenwich, Conn.: JAI Press.
Mossavar-Rahmani, Bijan. 1981. *Energy Policy in Iran: Domestic Choices and International Implications*. New York: Pergamon.
Mukherjee, S. K., and S. H. Rahman. 1982. "Energy-Economic Simulation Model for Oil Importing Developing Countries: An Application to India." Paper presented at the Fourth Annual North American Meeting of the IAEE.
Myers, John G. 1975*a*. "Energy Conservation and Economic Growth: Are They Incompatible?" *Conference Board Record* 12 (2).
Myers, John G. 1975*b*. "Oil Imports, Energy Conservation and Economic Growth." Paper presented at the Conference on Economics of Scarcity.
Myers, John G., and others. 1976. "Impact of OPEC, FEA, EPA, and OSHA on Productivity and Growth." *Conference Board Record* 13 (4).
National Academy of Sciences. 1980. *Proceedings of the International Workshop on Energy Survey Methodologies for Developing Countries, 1980*: Washington, D.C.
National Research Council Committee on Nuclear and Alternative Energy Systems. 1979. *Energy in Transition, 1985-2010: Final Report*. San Francisco: W. H. Freeman.
Nehring, Richard. 1978. *Giant Oil Fields and World Oil Resources*. Santa Monica, Calif.: Rand.
Nordhaus, W. D. 1973. "The Allocation of Energy Resources." *Brookings Papers on Economic Activity* 3.
Nordhaus, W. D. 1979. *The Efficient Use of Energy Resources*. New Haven, Conn.: Yale University Press.
Nordhaus, W. D. 1980. "The Energy Crisis and Macro Economic Policy." *Energy Journal* 1(1).
Odell, Peter R. 1970. *Oil and World Power: A Geographical Interpretation*. New York: Taplinger.
Odell, P. R., and K. E. Rosing. 1980. *The Future of Oil: A Simulation Study of the Interrelationship of Resources, Reserves, and Use, 1980-2000*. London: Kogan Page.
Pachauri, R. K. 1977. *Energy and Economic Development in India*. New York: Praeger.
Pachauri, Rajendra K., ed. 1980*a*. *Energy Policy in India: An Interdisciplinary Analysis*. Bombay: Macmillan.

Pachauri, R. K., ed. 1980*b*. *International Energy Studies*. New York: John Wiley.
Pachauri, R. K. 1982. "Financing the Energy Needs of Developing Countries." *Annual Review of Energy* 7.
Parikh, Jyoti K. 1980. *Energy Systems and Development: Constraints, Demand and Supply of Energy for Developing Regions*. New Delhi: Oxford University Press.
Park, Yoon S. 1976. *Oil Money and the World Economy*. Boulder, Colo.: Westview Press.
Penrose, Edith. 1968. *The Large International Firm in Developing Countries: The International Petroleum Industry*. Cambridge, Mass.: MIT Press.
Ponitz, Erwin. 1978. *Residential Energy Use Model for Austria*. Laxenburg, Austria: International Institute for Applied Systems Analysis.
Prast, William G., and Howard L. Lax. 1983. *Oil Futures Markets: An Introduction*. Lexington, Mass.: Lexington Books.
Reddy, A. K. N. 1978. "Energy Options in the Third World." *Bulletin of Atomic Scientists*, March.
Rouhani, F. 1971. *A History of OPEC*. New York: Praeger.
Sampson, A. 1975. *The Seven Sisters: The Great Oil Companies and the World They Made*. New York: Viking.
Schliephake, Konrad. 1977. *Oil and Regional Development: Examples from Algeria and Tunisia*. New York: Praeger.
Schurr, Sam, and Bruce Netschart. 1960. *Energy in the American Economy, 1850–1975*. Baltimore: Johns Hopkins University Press for Resources for the Future.
Searl, M., ed. 1973. "Energy Modelling." Working Paper. Washington, D. C.: Resources for the Future.
Seymour, Ian. 1980. *OPEC: Instrument of Change*. New York: St. Martins Press.
Simmons, Andre. 1981. *Arab Foreign Aid*. London: Associated University Press.
Smil, Vaclav. 1976. *China's Energy: Achievements, Problems, Prospects*. New York: Praeger.
Stewart, Hugh B. 1981. *Transitional Energy Policy, 1980–2030: Alternative Nuclear Technologies*. New York: Pergamon.
Stobaugh, Robert, and Daniel Yergin. eds. 1979. *Energy Future*. New York: Random House.
Stock, Francine. 1982. "Slow Progress on Liquids from Coal." *Petroleum Economist* 49 (2).
Stock, Francine. 1982. "Gulf States Face New Strains." *Petroleum Economist*, 49(7).
Streeton, Paul, and Shahid Burke. 1975. "Basic Needs: Some Issues." *World Development* 6 (3).
Sweeney, James L. 1977. "Economics of Depletable Resources: Market Forces and Intertemporal Bias." *Review of Economic Studies* 44, February.
Taher, Abdulhady Hassan. 1982. *Energy, a Global Outlook: The Case for Effective International Cooperation*. New York: Pergamon.

Tempest, Paul, ed. 1982. *International Energy Markets*. London: Oelgeschlager, Gunn, and Hain.
Thomas, John A. G., ed. 1977. *Energy Analysis*. Guildford, Surrey: IPC Science and Technology Press.
Turner, Louis. 1982. "OPEC Joins North-South Debate." *Energy Policy* 10 (2).
Tyner, Wallace Edward. 1978. *Energy Resources and Economic Development in India*. Bombay: Allied Publishers.
United Nations Conference on Trade and Development. 1980. *Energy Supplies for Developing Countries: Issues in Transfer and Development of Technologies*. New York.
United Nations Industrial Development Organization. 1979. *International Financial Flows: An Overview in Industry 2000*. Vol. 1 *New Perspectives*. Vienna.
United Nations Industrial Development Organization. 1982. *Impact of Higher Energy Prices on the Industrialization of Developing Countries with Special Reference to Least Developed Countries*. Vienna.
University of Oklahoma, 1976. "Our Energy Future: The Role of Research, Development and Demonstration in Reaching a National Consensus on Energy Supply." Report NSF/RA/760158. Norman, Okla.
U. S. Department of Energy. 1978. "Examination of Several World Oil Price Scenarios Given Exogenous Supply and Demand Estimates." DOE/EIA-0102/9. Washington, D.C.
U.S. Department of Energy. 1981. "Energy Projections to the Year 2000." DOE/PE-0029, Washington, D.C.
U. S. Federal Energy Administration, 1975. "Energy and Economic Impacts of Alternative Oil Pricing Policies." Washington, D.C.: Office of Data Analysis.
Verleger, P. 1982. *Oil Markets in Turmoil*. Cambridge, Mass.: Ballinger.
Weinberg, Alvin M. 1959. "Energy in the Ultimate Raw Material." *Physics Today*.
Workshop on Alternative Energy Strategies. 1977. *Energy: Global Prospects 1985–2000*. New York: McGraw-Hill.
World Bank. 1980a. *Energy in the Developing Countries*. Washington, D.C.
World Bank. 1980b. *World Development Report, 1980*. Washington, D.C.
World Bank. 1981a. *Global Energy Prospects*. Washington, D.C.
World Bank. 1981b. *World Development Report, 1981*. Washington, D.C.
World Bank. 1981c. Development Prospects of the Capital-Surplus Oil Exporting Countries." Staff Working Paper 483. Washington, D.C.
World Coal Study. 1980. *Coal: Bridge to the Future*. Cambridge, Mass.: Ballinger.
World Energy Conference, Oil Substitution Task Force. 1983. *Oil Substitution: World Outlook to 2020*. New Delhi: Oxford University Press.
Wright, Claudia. 1979. "Iraq: New Power in the Middle East." *Foreign Affairs* 58 (2).
Zenoff, David B., and Geoffrey S. Howard. 1981. "LDC Debt: Is the Sky Falling?" *Best of Business* 3 (2).

# INDEX

Al-Nar oil refinery, 87
Anglo-Persian Oil Company, 54–55
Arab League, 57
Arab Petroleum Congress, 57

Ba'athist regime, 82
Bank for International Settlements, 94–95, 112
"Bean counting" approach, 72
Blinkers-on scenario, 49–52, 149, 153–58, 165, 168, 175
Brandt Commission, 120–21, 129–33, 141, 172, 174
British Petroleum, 54
Brookhaven National Laboratory, 138
Brookings Institution, 31
Buffer stocks, 119, 120, 127
Burmah-Shell, 54

Camp David, 70
Cancun Summit, 127
Cartels, 42, 61–63, 95, 170–71, 173
Ceilings on oil production, 72–73
Centrally planned economies, 17, 29, 36–37, 43, 52, 80
Charte de l'Organisation Socialiste des Enterprises, 79–80
Charter, OAPEC, 70
Charter, OPEC, 70–71
Charter of Economic Rights and Duties of States, 117
Coal, 43–44
Constrained abundance, 48–49

D'Arcy, W.K., 54
Dhofar, civil war in, 88

Economies of scale, 53–54
Elasticity control parameter, 35
Energy, new and renewable forms of, 26, 28, 29

Energy consumption, 15–19
Energy elasticity parameter, 35–36
Energy policies, 30–52
Energy price, 14–15
Energy supply, 19–29, 42–52
Energy transition, 12–29
Eurodollar market, 80, 114
Exosmatic organs, 17–18
Exxon Corporation, 42, 47, 54

Fahrid, oil field at, 88
Faisal (Crown Prince), 85
Ford Foundation, 30, 31
French policy on nuclear power, 45

Geothermal power, 21, 44
Global welfare state, 132
Group of Seventy-seven, 130
Group of Twenty-four, 130
Gulf (oil company), 54
Gulf Cooperation Council, 75

Historical development of OPEC, 58–61, 77–88
  Algeria, 77–80
  Bahrain, 87–88
  Iran, 80–81
  Iraq, 81–82
  Kuwait, 82–83
  Libya, 84–85
  Oman, 88
  Qatar, 87–88
  Saudi Arabia, 85–86
  United Arab Emirates, 87–88
Horizontal growth of oil industry, 53–54
Hydropower, 21, 44

Institute of Energy Analysis, 35
Integrated Programme for Commodities (UNCTAD), 118, 119–20

189

## 190 / Index

International Association of Energy Economists (IAEE), 2
International economic order, 94, 115–28, 130, 132, 147, 177
International Institute of Applied Systems Analysis (IIASA), 30
International Monetary Fund, 113, 117, 127, 130, 141, 171
Iranian revolution, 26, 63–64, 65, 80
Iran-Iraq war, 69, 81, 82, 154
Irreversible conservation, 3
Islamic brotherhood, OPEC and, 101

Kuwait Fund for Arab Economic Development, 80

Libya, 58, 60, 69, 84–85
Licences to oil companies in Venezuela, 57
London interbank offered rate (Libor), 133

Marshall Plan, 98
Massachusetts Institute of Technology, 140
Mobil, 54

Natural gas, 49, 50
Nehring, Richard, 47
Neoclassical view of OPEC, 72
Noncommercial energy consumption, 15
North-South relations, 8, 115–17, 120-21, 127–28, 129, 147
Nuclear power, 19, 21, 44, 45

Oak Ridge Associated Universities, 35
Occidental Petroleum, 58–59
Odell, Peter, 46–47
Oil boycott, 5, 60
Oil embargo (1973), 5, 60, 94
Oil industry, history, 53–61
Oil market power, shift in, 5–11
Oil price revolution, 56–57
Oil price shock (1973–74 and 1979–80), 4, 94–96
Oil pricing system, 56
Oil production quotas, 71
Oil scarcity syndrome, 46
Organisation for Economic Cooperation and Development (OECD), 1, 26
Organisation of Arab Petroleum Exporting Countries (OAPEC), 58, 60, 70, 74, 143

40–42, 80, 98–114, 136–37, 142–46, 170
Organisation of Petroleum Exporting Countries (OPEC), 1, 2, 5–11, 36, 40, 42, 48–52, 56–75, 94–114, 127–28, 137, 142–77
charter, 70–71
history of, *see* Historical development of OPEC
Import Price Index, 103
Multiplier, 2
Press release Number 2–75, 115; Number 9–77, 102; Number 6–78, 103
sensitivity to market changes, 66–67
stability, 67–69

Pearl fishing, 87
Persia, discovery of oil in, 54
Persian Gulf nations, 48–49, 55, 59, 172
Petrochemicals industry, Kuwait, 163–64
Price elasticity, 34–36
Protectionism, 126
Purchasing power parity index, 89–90

Qaddafi, Muammar, 58, 60, 84, 85
Qatar Petroleum Company, 88

Reagan administration, 121, 141
Regional development banks, 131
Recovery factor, 26
Resources, country's capacity to absorb, 161, 164–65
Resources for the Future, 31
Royal Dutch Petroleum Company, 54

Saud (King), 85
Saudi Arabia, influence on OPEC stability, 65–68, 74
Scarcity scenario, 159, 160, 161, 162, 163, 164, 165–68, 175–76
Seven sisters, 54
Shell Transport and Trading, 54
Shia fundamentalism in Iran, 75
Sixth Special Session of General Assembly, 117
Social Democrats, 131–32
Sonatrach, 79
Special drawing rights, 130–31
Standard Oil, 54
"Struggle for the world product," 147

Surface Mining Control and Reclamation Act (1977), 43

Tariff concessions, 126
Texaco, 54
"Triangle of suspicion," 10–11

United Nations Conference on Trade and Development (UNCTAD), 118–120, 124, 148
United Nations General Assembly, 117
United States Department of Energy, 37–41, 49, 169

*Wall Street Journal*, on Brandt Commission, 131–32
Williamsburg summit, 147–48
Workshop on Alternative Energy Strategies, 31
World Bank, 2, 141, 142, 171
 forecasts, 14, 36–37, 45, 52, 126, 138–39, 162
World Coal Study, 44
World development fund, 131